You Are What You Click

You Are What You Click

How Being Selective, Positive, and Creative Can Transform Your Social Media Experience

BRIAN A. PRIMACK, MD, PHD

You

Library of Congress Cataloging-in-Publication Data
Names: Primack, Brian A., author.
Title: You are what you click : how being selective, positive, and creative can transform your social media experience / Brian A. Primack, MD, PhD.
Identifiers: LCCN 2021023540 | ISBN 9781797203645 (hardcover) | ISBN 9781797203652 (paperback) | ISBN 9781797203669 (ebook)
Subjects: LCSH: Online social networks--Psychological aspects. | Social media--Psychological aspects. | Personality.
Classification: LCC HM742.5 .P75 2021 | DDC 302.30285--dc23
LC record available at https://lccn.loc.gov/2021023540.

Manufactured in the United States of America.

Design by Brooke Johnson.
Typesetting by Maureen Forys, Happenstance Type-O-Rama.
Typeset in Mercury Text, Aktiv Grotesk, and Tiempos Headline.

This book contains advice and information relating to health and interpersonal well-being. It is not intended to replace medical or psychotherapeutic advice and should be used to supplement rather than replace any needed care by your doctor or mental health professional. While all efforts have been made to ensure the accuracy of the information contained in this book as of the date of publication, the publisher and the author are not responsible for any adverse effects or consequences that may occur as a result of applying the methods suggested in this book.

10 9 8 7 6 5 4 3 2 1

Chronicle books and gifts are available at special quantity discounts to corporations, professional associations, literacy programs, and other organizations. For details and discount information, please contact our premiums department at corporatesales@chroniclebooks.com or at 1-800-759-0190.

CHRONICLE PRISM

Chronicle Prism is an imprint of Chronicle Books LLC,
680 Second Street, San Francisco, California 94107
www.chronicleprism.com

For my parents, my children, and my soul mate

Table of Contents

Part 2

Be Selective

Part 3

Be Positive

Part 4

Be Creative 127

Introduction

IT WAS A SATURDAY MORNING in Pittsburgh—October 27, 2018—and I was making bagels from scratch. Outside was a crisp but sunny day, and I had somehow convinced the kids—ages eleven and fourteen at the time—to stay home and help me make brunch for some friends who were visiting from out of town.

I know what you're thinking. Why bother to make bagels from scratch?

Even though we easily could have strolled up the street to an Einstein's—they have the complex kneading and boiling process down—shaping the dough into floppy rings with my kids had this remarkable way of slowing down time. It gave us all a moment to pause from our phones, and it signaled the transition out of a busy week. Though, in today's world, our phones never stay silent for long.

Ding. When the first text came through, we didn't think much of it. Receiving a vague message about staying inside if you were in Squirrel Hill—our Pittsburgh neighborhood—wasn't particularly alarming. I kept kneading.

But a few minutes later, another *ding.* As my wife looked down at her phone, I noticed her face contort with tears. I ran over. "What?" I asked quietly. She shook her head and swallowed. She was barely able to mouth the words: "Tree of Life. Seven dead so far."

Tree of Life was the building where our Jewish congregation called Dor Hadash met. We might have been there for that Saturday's morning service had we not been preparing brunch for our guests. My wife and daughter had been there the prior week. It's a lay-led congregation, so I led services and study groups frequently.

A gunman had entered the synagogue, shouted, "All Jews must die," and opened fire. By the end of the rampage, eleven were dead and several were severely wounded, including law enforcement officials who were finally able to subdue the attacker. The gunman was brought to a local hospital and cared for by Jewish doctors.

The next few weeks of our lives—as a family and as a congregation—were focused on acute management of the situation. Funerals for the dead. Hospital visits for those who were critically injured. Vigils and meetings. Supporting the families of the deceased. Helping the traumatized.

As things started to go back to "normal," I found myself frequently returning to the question of what role social media had played in these events. As a doctor and researcher, I study the relationships among media, technology, and health. Now, this tragic incident brought these issues to me in a way I had never experienced so immediately.

It quickly became clear from news reports that the shooter had leveraged a particular social media platform, Gab, to gain support and encouragement for his anti-Semitic views and violent tendencies. The last thing he did before beginning his rampage was to send a message to his followers on that platform, saying, "I'm going in."

Of course, we will never know what this person would have done if he hadn't had support from people on that platform, but an argument can be made that his experiences on social media facilitated the incident.

On the other hand, as my community navigated its immense grief, social media became a crucial avenue for healing. Resources and support poured in from thousands of people all over the world. The sharing of information over social media led to remarkable acts of generosity and healing. For example, within days, the Muslim community raised enough funds to pay for all eleven funerals. My family and other members of our congregation received beautiful, heartfelt messages from people we had not spoken with in years—and from others we did not know—from across the globe. Would this compassion have been transmitted to this extent without social media?

Obviously, tragedies like this are shaped by many, many factors, including the availability of deadly weapons, hateful rhetoric from societal leaders, and the ongoing impact of complex historical events. But today's media and technology can powerfully amplify and facilitate the dangerous messages and beliefs that lead to violent actions.

A NEW APPROACH TO DIGITAL WELLNESS

For the better part of two decades, I considered writing a book about our relationship with social media and digital technology. The desire intensified whenever my research team came out with a new study on a risk or benefit of technology—or when one of my patients described how technologies like these have influenced their health.

But after the Tree of Life shooting, I felt an increased urgency. This tragic event and the outpouring of support in the aftermath showed me—with a clarity I never had before, even after twenty years of research—how social media and related technologies have become the sharpest double-edged sword of our era. On a daily basis, social media can catalyze connection, warmth, and generosity. On the other hand, it can breed feelings of depression, isolation, anxiety, and hatred.

More than ever, everyone—from children to young adults to the elderly—must deal with the consequences of painful tradeoffs related to social media. Because the way we use social media can negatively affect our health, many people have felt that the only solution is to stay offline as much as possible, to delete the apps, and to live life in the moment without constantly recording it.

But for most people, complete digital abstinence is not an option. We work with social media. We rely on social media for information and connection. We also realize that social media platforms—despite their challenges and drawbacks—are remarkable innovations that *can* improve our lives if we use them in the right way.

Take, for example, how technology benefited so many people during the 2020 coronavirus pandemic. Covid-19 thrust our world—quite suddenly—into previously unknown territory. Many people were directly affected: They lost their jobs, loved ones, or both. Even those not directly affected experienced dramatic shifts in their routines, along with increased confusion, uncertainty, and anxiety.

Technology softened the blow for many. People created virtual gatherings to celebrate graduations, birthdays, and religious events. Others used social-connection software like Zoom, Slack, and Teams—which they had only vaguely heard of before—to connect with friends through virtual coffee dates, book clubs, happy hours, and knitting clubs. Despite the many restrictions people faced during the global pandemic, technology enabled them to continue their jobs, educational pursuits, and other passions. Beyond the crisis, many of these connections and digital activities will continue.

But how do we balance the challenges of technology with the benefits?

That's where this book comes in. We need practical, evidence-based suggestions on how to maximize the value of social media in our lives while minimizing—as much as possible—the potential drawbacks. The goal of this book is to help you live in *balance* with social media.

Think of it like food. There's always a new study coming out warning us about something unhealthy in our diet: We should avoid too many carbs, too much fat, too much of a specific kind of fat, too much processed sugar, too many pesticides. To protect and improve our health, we must constantly adjust what we eat. But the answer, of course, isn't to stop eating entirely. Instead, we use the evidence we have to create a model—often represented by the "food pyramid"—to guide our dietary choices.

The common phrase "you are what you eat" reflects an inherent truth: The nutrients we take in basically *become* our bodies, so the healthier our diets are, the healthier our bodies will be.

The same is true with our social media consumption and our use of other digital technologies. We can and should create a healthy technology "diet," one that improves, rather than harms, our bodies and minds. What we consume online—whether through scrolling, listening, watching, or clicking—fills our minds with "scripts" about ourselves, others, and the world. This changes what we think about and how we feel. These changes in our minds can affect us physically—by influencing the stress hormones flowing through our arteries, our blood pressure, and even whether we get diabetes or cancer. In other words, we become what we click.

Because media and technology are relatively new phenomena, we have no natural immunity to their negative impacts, which can go unchecked. Millions of years of evolution have ensured that our bodies are equipped to process out toxins, heal infections, and defend against microbes. We don't have innate filters like these for tech. Instead, our minds and bodies act like sponges, absorbing anything and everything we immerse ourselves in digitally. This is why, as our use of and reliance on technology increases, we need a set of guidelines—like the food pyramid—to help us craft a satisfying and nourishing tech diet.

A few things are necessary to create an effective digital wellness plan. First, the plan needs to be simple. We can't remember or implement a laundry list of 126 things to do and avoid. This is why, when it comes to nutrition, the US Department of Agriculture created the food pyramid in the first place. It's also why this was eventually replaced with the even simpler "MyPlate," which reduces thousands of nutritional choices into five basic boxes.

While social media and digital technologies are extremely complex, we can define simple principles to guide our digital wellness. Over the course of my career, I have always come back to three major ideas that I believe can transform our relationship with social media and digital technologies: *Be selective, be positive, and be creative*. This book unpacks each of these principles as part of a "social media pyramid" and shows how to apply them to specific challenges in our digital and online worlds.

How can we feel truly *connected* and *heard* when communicating remotely? How can we get the information and news we seek without plunging down a rabbit hole of doomscrolling? How can we optimize *when* and *how* we use social media so we feel empowered and not overwhelmed? These and many more questions will be answered here.

Another thing to keep in mind is that tech diets need to be *individualized*. Each person experiences and responds to technology in their own way. Each person is exposed to their own special mix of messages, advertisements, platforms, and feeds, and each person's unique personality and circumstances can have a dramatic effect on how social media impacts them.

Some people are crushed if they are "unfriended" on social media, sending them spiraling into self-doubt, while others are pleased to have their friend list automatically curated for them in this way. Some people breathe a sigh of relief when they find out that their meeting will be over Zoom, while others intensely crave in-person experiences and dread the awkwardness of a video chat. This book will help you individualize your experience based on your personality. It will also show you how to curate what you consume online based on what brings you more joy, inspiration, and peace of mind.

Finally, this book will offer a *positive and proactive approach* that is missing from a lot of advice about technology.

Early nutrition recommendations were usually phrased in the negative, such as, "Don't eat processed foods, carbs, or red meat." But now eating advice is more frequently put in positive terms. We are told what *to* eat, rather than what *not* to eat. Instead of a long list of no-nos, diet advice focuses on the importance of having enough fiber, protein, and fresh fruits and vegetables.

This is important. According to behavioral scientists, people are more likely to follow suggestions that are positively framed. Instead of feeling shame and guilt over social media use, we should develop better practices that foster empowerment and positivity. Any technology can be misused, but the tools in this book will help digital tech work for you instead of against you.

It's time we were empowered to use social media and digital technology to our advantage. They are here to stay, so let's use them wisely and follow a realistic, positive model for creating lifelong digital wellness. The model I propose in this book is simple but flexible; anyone can use it while adapting it to their individual personality and needs. And though it acknowledges negative impacts of social media, it ultimately focuses our time and energy on creating positive experiences and feelings. Let's begin.

Part 1

Why We Need a "Food Pyramid" for Social Media

SOCIAL MEDIA ISN'T THE FIRST technology in human history to disrupt our lives in a major way. The printing press, telephone, radio, television, and microchip have all dramatically changed society.

Yet many scholars agree that the rapid adoption of social media is unlike anything we've seen before. While the term itself is only a couple of decades old, billions of people across the world use social media for hours a day.

In part, this is due to how strongly linked this technology is to emotion, both negative and positive. There is no shortage of scientific studies linking social media use to mental health issues, like depression, anxiety, and loneliness—and each of these conditions is currently at epidemic levels across the globe.

When used well, however, social media can facilitate joy, comfort, and friendship. So, pulling the plug on this technology is not just unfeasible in today's world—it also could result in squandering a lot of potential benefits.

Social media is powerful because it is personal in ways that previous technologies are not. The printing press didn't follow you around, learn from your behaviors, and use that information to influence you. Television didn't substantially leverage your relationships with your friends, family, and acquaintances to give itself greater impact in your life.

Beyond being personal and personalized, we also can't seem to stop looking, scrolling, or clicking because of the wide spectrum of feelings social media evokes.

We've had fabulous experiences on social media of feeling valued and included. We've marveled at how easy it is to stay in touch with people we had thought we might never see again, and we've found information and content that delighted us or changed our perspectives.

But we've also been disappointed by social media. To lesser or greater degrees, we've had experiences that made us feel marginalized, belittled, or misunderstood. We've also had concerns about our privacy. This combination of good and bad experiences often sets up a classic yo-yo behavior—a cycle of ups and downs that ends up leaving us feeling more alone, anxious, and stressed out.

We need an empowering system that keeps us at the center of the experience and helps us live in balance with this remarkable technology. We need a system that's flexible so that it will work across platforms and across time. We also need something simple—a plan we can easily incorporate into our already very full and busy lives. And we need a system that *works*. I created the social media pyramid to check off each of these boxes.

But to understand how the pyramid works—and to immediately put it to use in our lives—we first need to explore why it's so needed. Our story starts in London.

1

The Minister of Loneliness

IN 2018, THE UK PRIME minister created a new position for a "Minister of Loneliness." Usual ministerial posts are for things like foreign affairs, housing, and education. The last time a cabinet post in the United States was created was for homeland security.

Are loneliness and emotional health problems in the world really *that* bad?

In a word, yes.

Over the past couple of decades, loneliness has become a worldwide epidemic. In the United States, 61 percent of people feel lonely—not just now and then, but on a regular basis. In many populations, people consider themselves closer to their television or a pet than to other people. Don't get me wrong—I love my pets. My dog and two guinea pigs bring me joy. But something is off-kilter when we connect more to a television than to our own family and friends.

We all understand that loneliness is emotionally painful, but it also can be truly damaging—and even deadly.

A 2015 report involving millions of people found that those who were lonely had a 26 percent increase in the chance of dying compared with those who were not lonely. Similarly, people who lived alone had a 32 percent increased risk of dying compared with those who lived with others. This puts loneliness and living alone on par with other serious risk factors for death like heart problems, smoking, and obesity.

How can loneliness have this powerful an effect?

One reason is because loneliness is closely linked to more serious mental health conditions. Sure, everyone gets lonely at times, but prolonged loneliness can lead to other conditions like depression and anxiety. Both of these conditions can increase stress hormones that increase the risk of things like heart problems and cancer. Also, if you're isolated or living alone, you don't have as many people to help you take your medication, to take a walk with, and to remind you to take care of yourself.

Other emotional and mental health problems are also on the rise. One-quarter of Americans rate their own mental health as "fair" or "poor," and this number has increased significantly even in the past decade.

In 2016, the *New York Times* reported that the US suicide rate had surged to a thirty-year high. Less than a year later, the World Health Organization declared that emotional health problems had replaced muscle and bone pain as the leading cause of *disability* worldwide.

What? The number of people who can't work because they are too depressed or anxious has become higher than the number of people who can't work because of injury, heart disease, cancer, or diabetes? Again, the answer is yes.

Emotional health issues are more ubiquitous than many people realize, because the millions of people who struggle with them don't just come up to you when they meet you and say, "Hi! I have severe depression!" Studies have shown that about 20 percent of Americans will become clinically depressed in their lifetime, but recent reports say it may be even more common than that.

For example, a 2017 study showed that about 27 percent of people who visit the doctor for routine things (like ankle sprains, back pain, or diabetes) also have significant depressive symptoms at the same time.

Emotional conditions like depression and anxiety impact our physical health in different ways. Depressed people are less likely to take their blood pressure medicines regularly, which sets them up for heart attacks and strokes. And if they do have a heart attack or stroke, depressed people have a harder time recovering compared with non-depressed people. People with anxiety are more likely to smoke, drink alcohol to excess, and do other things that seriously harm their health. Ultimately, this translates into significant loss of life.

SPOTLIGHT ON SOCIAL MEDIA

Depression and anxiety have always existed, but they weren't always *this* bad. Why does Britain suddenly need a Minister of Loneliness? Why are things like depression, anxiety, and suicide increasing—all over the world—while we're spending billions trying to identify, prevent, and treat emotional health conditions?

Some experts point to our increased reliance on digital technologies like social media. Psychologists, sociologists, and epidemiologists are connecting the rise in emotional health problems with the rise of social media. Psychologist Jean Twenge is one, and in books like *iGen: Why Today's Super-Connected Kids Are Growing Up Less Rebellious, More Tolerant, Less Happy—and Completely Unprepared for Adulthood—and What That Means for the Rest of Us*, she demonstrates how increased use of social media and related technologies seems to parallel increases in emotional health problems among youth. She also argues that impersonal technologies such as social media may be replacing more beneficial face-to-face social activities.

One reason social media may have such an effect is because of how relatively new it is; we're still learning how to use social media and not have it use us. The term *social media* only originated around 2004—and yet by 2020 about four billion individuals used social media worldwide. So the number of social media users on Earth increased from zero to about four billion in a couple of decades. This makes social media one of the most rapidly adopted—if not *the* most rapidly adopted—technology in history.

Today, the most commonly used social media platform is still Facebook. But as we all know, there are many other platforms, like Twitter, Instagram, TikTok, Snapchat, and Reddit. Facebook currently has between two and three billion users, if you define a "user" as someone who has logged on in the past month. This is more than seven times the population of the United States. In January 2018, TikTok had about fifty million users—a pittance compared with the eight hundred million users it had by 2020.

Most young adults spend two to four hours per day on social media. But the fastest-growing population of users are adults sixty-five and over. In the past decade their use has increased by more than a factor of five.

Then again, we're basically all users of social media now, even if we don't think of ourselves that way. The *Cambridge Dictionary* defines social media as "websites and computer programs that allow people to communicate and share information on the internet using a computer or cell phone." Who *doesn't* do this? We are using social media when we join a Zoom call and when we watch videos on YouTube, even if we don't spend time reading or posting comments.

Video games are almost all classified as social media these days. In the 1980s, arcade games like *Frogger*, *Pac-Man*, and *Tempest* were solitary. Sure, you might alternate playing with a friend, but that was as social as it got. Today, nearly every popular video game—such as *Civilization*, *Minecraft*, *Fortnite*, and *Grand Theft Auto*—involves interaction with people around the world. The interfaces make it easy to play and chat with those people—or obliterate them if you'd like.

What about texting? Some communications and marketing professionals debate whether texting "counts" as social media. However, there are good arguments for why today's texts qualify. First, using the *Cambridge Dictionary* definition, the texting app on your phone certainly allows people "to communicate and share information."

Second, texting apps are becoming more "social" over time. Texting originally involved mostly one-to-one communication, but newer apps make it easy to form and communicate with groups. Finally, texting and other social media platforms exist synergistically. For example, Facebook Messenger (a "texting" application) seamlessly integrates with Facebook.

Many other platforms also count as social media even if they don't seem like social media. You might go to Goodreads to find a book recommendation and suddenly start debating with someone across the world about the merits of the most recent Jodi Picoult novel. This interactivity is also why sites like the Internet Movie Database (IMDb), Nextdoor, and LinkedIn are functionally similar in many ways to other social networking sites.

So, we basically all use social media, and use is increasing. At the same time, there is an epidemic of loneliness and an increase in rates of depression and anxiety globally. But does that mean these two things are related? Some say yes, and some aren't sure, so it's worth looking more closely at the links and parallels.

2

Goldilocks Was Nowhere to Be Found

JUST BECAUSE TWO THINGS GO up and down at the same time doesn't mean that they are directly related. The classic example is that as ice cream sales go up, so do drownings.

Of course, eating more ice cream doesn't make people drown. Nor, when people drown, do others seek out more ice cream. Neither is *caused* by the other. Instead, both are affected by a third thing: the time of year. Both buying ice cream and drowning go up in the summer.

To determine cause and effect, you need to do good research. For example, you need to control for appropriate outside variables (like time of year), measure the variables as best you can, and be cautious when interpreting what you find.

So, back to our key question: Is the rise of social media related to the increase in emotional health issues? Or are there outside variables making them *appear* connected? My research group and I decided to put these questions to the test. We conducted a national study with about two thousand young adults. And what we found surprised us.

We predicted ahead of time that we would see what's sometimes called a "Goldilocks effect," a term based on the story of Goldilocks and the three bears. Goldilocks prefers a happy medium between two extremes, and we imagined the same thing must exist with social media use.

In this case, we guessed that, at very low levels of social media use, people might have increased depression and anxiety. This might be because they weren't getting the benefit of online interactions and connections. The same thing might also happen with very high levels of social media use: Too much time online might crowd out valuable "real life" experiences, which could lead to increased emotional health concerns.

The sweet spot might be at medium levels of use. Maybe getting a *moderate* amount of social media interaction would help people make connections, gather support, and feel that they were part of a community—but without negatively impacting other valuable in-person social and life experiences.

However, that was *not* the pattern we found. Instead, our study showed that every increase in social media use came with an increase in depression, and the same relationship existed for other negative outcomes we looked at, like anxiety, poor emotional support, and even social isolation.

Visualized as a graph, this relationship formed a straight, steep, upward line (see figure 1). Even when we took into account variables like sex, age, race, income, and living situation, we consistently found that people who used the most social media were about *three times more likely to be depressed* than people who used the least social media.

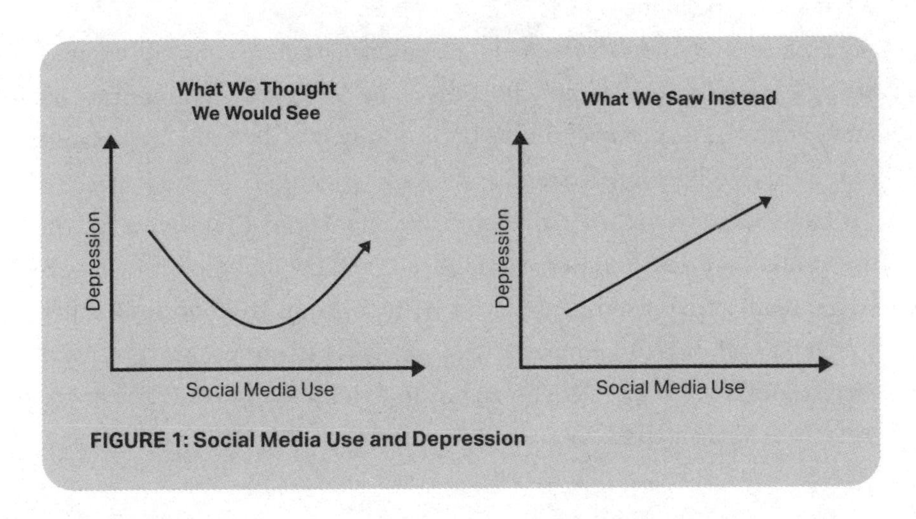

FIGURE 1: Social Media Use and Depression

Since that study was published, several others replicated these findings. These studies suggest that the rising emotional health epidemic and the increase in social media use are not simply a coincidence—they are somehow linked.

CHICKEN OR EGG?

What our study *didn't* tell us was which came first: social media use or depression. Do people turn to social media because they are depressed, or does social media lead to depression?

We imagined that there might be some truth to both directions. This is often the way it is in social science. Some people might self-medicate by using social media to manage existing emotional health concerns, suggesting that mental health problems *lead* to more social media use. On the other hand, healthy people who start to use more social media might *then* get more anxiety and depression for various reasons.

To figure out which came first, we had to follow people *over time*. So, we studied a thousand young adults who were *not* depressed at the beginning of the study and followed them for about six months. This enabled us to figure out who were the most likely to *become* depressed. Was it men or women? Was it richer people or poorer people? And what about social media use?

Interestingly, almost none of the standard demographics predicted who became depressed and who didn't. However, the amount of social media they used at the beginning of the study was dramatically related to who would become depressed after six months.

How strong was the relationship? Well, imagine that the people in the study were put into four equal-sized buckets based on how much social media they used. Compared with those in the bottom bucket (the least social media), those in the top bucket (the most social media) were about *three times* as likely to become depressed over the next six months. This is a huge difference.

Also, the relationship was linear—every increased level of social media use came along with more risk for becoming depressed—like what we see in figure 1.

What about the other direction? If you're depressed, does that lead to more social media use? To test this, we started with a mix of people who were depressed and others who weren't. We found that, compared with nondepressed people, depressed people *did not* start using more social media over time.

What these findings showed us is that social media seems to lead to depression, but depression does not seem to lead to more social media use.

So, if a large, national, well-controlled study shows that every increase in social media use is linked to a higher risk of becoming depressed, should we all just throw away our devices?

Certainly, some people might make that choice. But I personally don't think that it is the answer. The answer is that *we need to improve how we use this tool.* Study results like these suggest that, at this moment in time, we're not using social media very well—it's using us. But it doesn't *need* to be this way. Understanding the problem can be the first step on the journey to doing something proactive about it.

So, what could explain the links between social media use and poor emotional health, and what can we do about it?

3

A Social-Like Experience

ONE EXPLANATION FOR WHY SOCIAL MEDIA use might lead to emotional health concerns is that it may replace more valuable—and more "real"—in-person interactions.

In his book *In Defense of Food*, Michael Pollan uses the phrase "foodlike substances." Like food, foodlike substances stimulate taste buds and contain calories. The problem is that these foodlike substances don't give us many of the *nutrients* we need from food—and they often give us other things that we *don't* need, like trans fats, lots of sugar, and potentially harmful preservatives. Pollan describes how a lot of current health problems—such as epidemic obesity and diabetes—may be related to eating less actual food (such as fruits and vegetables) and more foodlike substances (like candy bars and soda).

In the same way, social media may be a "social-like experience." Yes, social media does involve social interactions. Yes, it conveys emotion. And yes, it often involves supportive communications, whether they are in the form of kind words, a "like," a thumbs-up, or even a 💩.

But just like Apple Jacks cereal doesn't provide the same nutritional value of actual apples, when it comes to life online, we're usually not getting the full emotional and psychological value of in-person social experiences. Our phones, the internet, and social media are designed to be sticky, to keep us coming back for more, and to create more craving than satisfaction. Sort of like sugar.

Why isn't social media an adequate substitute for the real thing? To answer that, we need to understand how important social contact is to us as a species.

A few years ago, my ten-year-old daughter came to my wife and me with a twenty-five-page PowerPoint presentation. I have no idea where she learned PowerPoint.

The presentation was all about sugar gliders—and why we desperately needed a sugar glider as a pet.

A sugar glider—which I'd heard about only vaguely—is a cute marsupial. It gets its name because it loves sugary foods like nectar. Its name also comes from the fact that it can glide through the air, sort of like a flying squirrel.

My daughter's PowerPoint featured pictures of adorable sugar gliders in various poses, along with a comprehensive list of reasons why sugar gliders make great pets. They don't tend to have medical problems. They can live a decade, so you can really bond with them. They love playing and cuddling in a warm pocket. They would help her learn a lot about biology, including what kind of food and how much activity they need.

My wife and I listened to the presentation with interest and a fair amount of amazement. Unfortunately, there were a few problems. One was that owning a sugar glider was illegal in Pennsylvania, where we lived at the time. They are also often taken illegally from Australia and sold on the black market.

Undeterred by our negative response, our daughter came back a couple of days later with a PowerPoint with twenty-five *more* slides about why we needed guinea pigs. They would teach her about responsibility and biology. She would feed them multiple times a day and clean their habitat herself. And one of the biggest guinea pig rescues in the country happened to be only twenty-five minutes from our home. So, could we at least just go and take a look?

That's all it took. We went a couple of days later, and we learned a tremendous amount about guinea pigs. The rescue received dozens of neglected guinea pigs from all over the country. The owner of the establishment, a fierce advocate for kindness to these animals, worked closely with a local veterinarian who specialized in exotic pets. Together, they nursed these remarkable creatures back to health and happiness and found them good homes.

We went home that day with two guinea pigs.

Given this emotionally poignant experience, and all we learned about how many of these creatures need good homes, it's probably not so surprising that we came home with one. *But two?*

The owner of the rescue explained to us that guinea pigs are such social creatures that, if you have just one, it tends to *die of grief.* Many people actually take three. That way, if one dies from an illness, there are still two left to take care of each other. I looked all of this up and confirmed what she said. That's how we ended up coming home with two.

It may seem dramatic to claim that a creature could "die from grief," but this is how it is with all social animals—*including humans.*

As I mentioned in chapter 1, well-designed studies have linked loneliness to an impressive list of mental and physical conditions. It can worsen blood pressure, increase the risk of heart disease, weaken the immune system, worsen the prognosis of cancer, and heighten the risk of dementia. When you crunch the numbers, loneliness is strongly linked to early death. Just like guinea pigs, we too can die of grief if we are lonely.

Can social media provide an antidote to this grief? It is *social,* after all. But it would be a mistake to simply equate social media with true in-person connection. Our species developed in person, together in community. The word *social* stands for all these things. Seeing someone smile. Interacting through physical activity. Hearing laughter and play.

I taught in West Africa after college, and I learned the appropriate custom to greet someone when passing by in local villages. You would stop, hold hands, and look into each other's eyes. You then would ask a series of questions in a ritualized order. How are you? How is your body? How is your family? The list went on. Then, only reluctantly, the ritual would come to an end as you both slowly moved away back toward your errands.

We all learned during the Covid-19 pandemic how critical true in-person connection can be. When we couldn't see people in person as frequently, we could at least achieve a certain level of closeness and satisfaction through digital platforms like FaceTime and Zoom. But many of us also realized how pale those experiences were in comparison to the real thing.

We evolved over eons to respond to a particular set of social cues, so it's not surprising that social media, as a modern replacement, feels insufficient. After all, comparing millions of years of being human to the last twenty years of social media is like comparing an entire lifetime to a fraction of a second.

My colleague Ariel Shensa and I recently published a study that looked into the difference between the value of face-to-face relationships and social media relationships. First, we discovered a dramatic association between *better* face-to-face emotional support and *less* risk of depression. The risk of depression went down 43 percent for every point on a five-point scale. That's a huge decrease, and it was consistent with prior work suggesting that good face-to-face support and less risk of depression go together.

Then came the interesting part. When we looked at the relationship between *social media–based* emotional support and depression, people who said they had *higher* social media–based emotional support had a significantly *increased* risk of depression!

These findings support the idea that what we feel from in-person relationships is somehow different from what we feel from social media relationships. Just like we evolved to need apples, not sugary cereal, we evolved to need true social experiences, and we need to be cautious about trying to replace them digitally.

SMILES VS. EMOJI

It's not that social media doesn't genuinely *try* to replace social interactions. There are online analogies for many physical things that are deeply social to us. A smile is an important social cue. So we have smiling emoji. Laughter is another important one, so it's no surprise that the most common emoji used represents a face laughing so much that it's crying. For this same reason, it makes sense that two of the most commonly used abbreviations in texts are LOL (laugh out loud) and ROTFL (rolling on the floor laughing).

The problem is that these are often insufficient comparisons to the real thing. In fact, I found it somewhat ironic when I recently noticed someone typing "LOL" and "ROTFL"—and adding a bunch of smiley emoji—while wearing a glum, sour expression. We can all relate to this. Next time you text someone ROTFL, observe yourself. Are you truly rolling around? Are you actually laughing, or even smiling?

Chapter 29 discusses how to use emoji to help online communications be more expressive and connecting. Yet despite the positive possibilities, there's a dramatic disconnect between an emoji representing a face and an actual face.

Consider, for example, that the human face has forty-three muscles. Most have fun, obscure names like orbicularis oculi, corrugator supercilii, zygomaticus minor, and levator labii superioris alaeque nasi. Theoretically, if we imagine every possible combination of these forty-three muscles, they could make 8,796,093,022,208 different facial expressions.

Clearly, our *real* face can communicate a whole lot more than a series of emoji, even if you select those emoji very carefully from a big list. This is only one example of how social media can't quite replace the real thing. Just like emoji only capture a small fraction of our true facial expressions, in-person physical, auditory, and other cues can be approximated through digital technology—but not replicated.

In today's world, however, every interaction simply *can't* be in person. Society is moving in ways that emphasize remote connections, and the ongoing impacts of the Covid-19 pandemic will likely accelerate and catalyze that move.

Importantly, though, the switch to more remote communication isn't just something to bemoan and live with. As later chapters discuss, there are ways of *leveraging* social media and other digital communications to be much more meaningful than they are today. Ultimately, we'll be best off if we *use* social media wisely instead of merely tolerating it. We need to incorporate a balance of in-person social interactions with innovative and meaningful social media use.

Social-like experiences alone won't stave off loneliness or quench our need to be social. An emoji just isn't the same as a smile. But this isn't the only reason using too much social media might lead to feeling down. There's also social comparison.

4

More Powerful Than Advertising

SOCIAL COMPARISON HAS ALWAYS BEEN a thing. It's hardwired into our brains. Psychologists have found, across numerous studies, that we don't just think about how we are doing in general—we tend to frame everything in terms of how we're doing *compared to everybody else*.

We all know how this plays out. We drive into work and someone has a fancier car. We walk into class and someone has more expensive designer jeans. We walk into someone's house and they have a Ming vase—and we don't. The biblical command "Thou shalt not covet" is there for a reason: It's deeply human to covet, and we do it all the time.

Social comparison is also deeply ingrained in other animals. For example, chimpanzees will happily accept just about any reward in any amount—grapes are a favorite—for doing various tasks in experiments. But chimps *watch one another*, and if one starts to get more grapes than the others for doing the same tasks, the maligned chimps will get upset, act out, or refuse to keep participating.

Animals also factor in how *popular* their peers are when they behave. In a 2017 French study, baboons were given a task to do on a computer monitor. On the monitor, a bunch of L-shaped images would appear, but among them was one T-shaped image. The baboons were taught that if they touched the T-shaped icon, they'd get a reward. When a baboon saw that a *more popular* baboon was doing the task correctly, they tended to try harder at the task. However, if a less-popular colleague was doing the task correctly, that didn't influence the baboons. In fact, it often made them less likely to try to do the task correctly.

And it's not just primates. Even guppies—yes, the fish!—compare themselves to other guppies. Male guppies try to mate with females who are being more aggressively courted by other males. Often, the male guppy will step up his game if he's more colorful than his competitors. In other words, the male guppy is always thinking and acting based on *how they compare to others*.

Even though social comparison has been around for eons, social media may be bringing our coveting to new heights.

One reason is that social media affords so many different possibilities for curating our lives. In person, it's harder to hide or maintain an ideal exterior. When we meet in daily life, we tend to notice imperfections and the real challenges people face. On social media, we can carefully choose one image out of hundreds to share with the world. Our day might be full of messy, difficult experiences, mistakes, frustrations, and unattractive moments—but we can still post about the one moment that makes us seem like we have it all together.

The second reason that social media is particularly powerful as an envy machine is this: We are observing *real people*—the ones we are hardwired to compare ourselves to.

Like baboons and guppies, we tend to compare ourselves to those who are more *like us*—in terms of background, experience, demographic characteristics, and other attributes. So, when I think about my financial situation, I don't compare myself to Bill Gates or Warren Buffett. Instead, I instinctively compare myself to other academics like myself.

Similarly, when it comes to athletics, I don't compare myself to Aaron Rodgers of the Green Bay Packers and wish that I too had an average quarterback rating over 100 in fifteen NFL seasons. Instead, I'm instinctively more likely to be envious of a friend's better 5K time or a work colleague's regular exercise routine.

This is what makes social media the ultimate breeding ground for envy. I'm not Facebook friends with Aaron Rodgers, Bill Gates, or Warren Buffett, but I *am* friends with hundreds of people who share my exact demographics. Every day, each can tell me the best thing that happened to them, their biggest achievement, the most impressive thing that their kid has ever done, and the most wonderful thing they just did for the environment. And I instinctively compare myself to all of them.

In this way, social media may be *more powerful and effective than advertising* at producing envy. And advertising is extremely powerful. One of its main functions is to make us feel like we're not good enough—unless we purchase a given car, computer, or brand of detergent. Worldwide advertising spending is now over half a trillion dollars. That's more than the gross domestic product of over 180 nations. Companies wouldn't invest this amount of cash if ads didn't work extremely well.

Advertisers use remarkable technologies and marketing concepts to capture our attention. One day our life is just fine the way it is, and the next we have an overwhelming urge to buy new headphones, a different brand of bottled water, or a digital device we can tell to turn the lights on so that we don't have to touch anything. Advertisers can create these urges in us consciously and subconsciously.

Yet despite this power, we recognize on some level that an advertisement is an advertisement. We understand that we are being sold to. We know ads are crafted to target our desires and showcase a perfect picture of what we want. We're aware that no one is as flawless and blissful as the actors in commercials, who have been aggressively made up, airbrushed, and sometimes surgically enhanced.

Social media, though, is different. Because social media is used by *real* people, most of whom you *know*, and who *haven't* been paid to put on a show, we accept the ways they present themselves as "reality."

Combine that with readily available tools like filters and Photoshop and you can see why social media can be so influential. Through it, anyone can curate themselves into an "advertisement" for the perfect life. On top of that, these people are peers and friends. Until now, very little has captured this dangerous combination.

THINK HORIZONTALLY

Social comparison can profoundly influence mental health. We don't usually compare ourselves negatively to someone, just shrug our shoulders, and move on. Teddy Roosevelt once stated, "Comparison is the thief of joy," and this has been supported by research.

The criteria psychiatrists use to diagnose mental conditions like depression include things like "feelings of worthlessness or excessive guilt," which is what constant social comparison can lead to. We ask ourselves: *What am I "worth"? Why aren't I as good as other people? Am I not as smart, talented, or hardworking? Am I a failure?* With social media, the tragic thing is that these thoughts are often based on inaccurate assumptions about people or the images they post.

One reason we're susceptible to this is that our brains have built-in hierarchies. We tend to think "vertically" and not "horizontally." The chimps and baboons we met at the beginning of this chapter innately know exactly where they stand in their groups. Evolutionarily, vertical thinking developed because it leads to a more orderly tribe. There isn't constant fighting and strife—and subsequent destruction of the tribe—because each individual knows where they stand. Of course, hierarchies can change. Now and then an individual challenges another "above" them, and if the challenger loses, the current hierarchy is reinforced. If they win, a new hierarchy is established.

As humans, we experience these hierarchies in similar ways. Our brains cling to things that are easily quantified, which is why many people use money as an indicator of status. We have extensive hierarchies in the working world. And, even if they're not delineated in corporate organizational charts, similar hierarchies also underlie our social and romantic relationships.

Social media did not invent social comparison or hierarchy. Those things have been with humans from the start. But there isn't much question that all the rosy, curated, glamorous portrayals we see on social media can make us feel bad in comparison, sometimes to clinical extremes.

The good news is that there is hope. There *are* ways of keeping social comparison at bay. Despite our hardwired awareness of where we stand within a hierarchy, we can learn to think *horizontally* on social media instead of vertically. With horizontal thinking, we can avoid comparing the accomplishments of others to ourselves in ways that make us feel bad; we can learn to feel happy for our friends *and* for ourselves (which chapter 27 discusses).

Social comparison is how overly *positive* messages on social media can paradoxically end up worsening our mood, but we get it from both sides. Overly *negative* messages can make us feel bad, too.

5

The Mean World Syndrome

LIKE MOST MEDIA, SOCIAL MEDIA is about the extremes: the highest highs and the lowest lows. Social comparison focuses on the highs. This chapter focuses on the lows.

What there *isn't* on social media is a whole lot about people's normal, humdrum routine. People don't generally post things like: "I just went to the bathroom"; "The dishes were pretty dirty today"; or "Brushing my teeth hurt more than usual tonight on the back left side, so I think I may need to see my dentist." But those are the kinds of things we actually experience every day.

Ironically, on social media, you're more likely to see the once-in-a-lifetime: "So proud that Chip just graduated *magna cum laude* from Harvard!"; "I just qualified for the Boston Marathon!"; or "Happy tenth anniversary to my soul mate—every moment has been sheer bliss!" These moments are genuine, but they omit or gloss over the pain, struggle, and behind-the-scenes drama that were also involved.

Our favorite medical dramas, the ones we can't seem to stop binge-ing, focus on just that—*drama*—and not what an emergency room is actually like. The same is true for legal, police, and crime dramas. I remember being surprised during my first emergency-department rotation in medical school. I was expecting a barrage of people running in with stretchers while teams frantically placed multiple IVs and breathing tubes—all while a full symphony orchestra played dramatic theme music.

The reality was quite boring. There were hours of waiting. Mesmerizing beeps came from each room, and almost none of those beeps signaled actual emergencies. They indicated things like the fact that Ms. Johnson had shifted in bed, and her wires needed to be adjusted.

When the medical drama *ER* premiered, applications to emergency-department residencies skyrocketed. However, many of those people eventually became disenchanted and switched careers.

Of course, these TV programs are entertainment. They *shouldn't* reflect reality. No one wants to watch a "realistic" crime drama that shows people sitting around drinking coffee and filling out paperwork. That wouldn't do so well in the ratings. Movies and TV focus on extremes for a reason.

But experiencing extreme media content comes with its own dangers, especially when it comes to the lowlights.

However, before exploring how sad, angry, and heavy social media content can affect our mood and mental health, I want to make clear that my focus is on the *personal impact* of this content, not on the content itself. I don't mean to suggest that negative material should not be shared. In the same way that we want our friends to celebrate their good news on social media, we can be grateful when they share their bad news, especially when it's news we might not hear another way.

In other words, there is *nothing wrong* with posting tragic news or feelings of anger and frustration on social media. In many ways, social media can be an *ideal* vehicle for delivering difficult news to a wide group of friends, coworkers, and distant relatives. When a person is battling cancer or mourning their loved ones, there are many physical and emotional demands competing for that person's time, and social media can help people share information efficiently. Moreover, it can help those affected get the support and compassion they so badly need from their community.

During the 2020 Covid-19 pandemic, when people had to remain quarantined and were unable to visit loved ones in the hospital or who were struggling with isolation at home, social media became a genuine lifeline. Social media also provides an outlet for expressions of anger and sadness over social issues, current events, and politics. At times, social media provides an important—even critical—way to share thoughts and provide comfort in the aftermath of tragic events like shootings and natural disasters.

We *need* outlets for conversations about current social issues. We *need* to find comfort in communities who will listen and understand our struggles. And we *need* people to mourn with us when there has been a death or other tragedy.

At the same time, *we need to be aware* of how powerful being exposed to such a high concentration of anger, sadness, and other painful emotions can be.

AS POWERFUL AS THE REAL THING

After the tragic events of September 11, 2001, when terrorists hijacked four planes and destroyed the World Trade Center, a study published in the *Journal of the American Medical Association* reported the results of a national survey looking at which people ended up getting posttraumatic stress disorder. The researchers surveyed people who were directly affected by the attacks—such as people who had been living in Washington, DC, and New York City. They surveyed people who were in the armed forces and the families of people who had been injured or killed. They also surveyed people who had no direct connection to the events. Who had the most posttraumatic stress disorder?

Interestingly, the researchers found that one of the strongest risk factors for ending up with posttraumatic stress disorder—if not *the* strongest risk factor—was *the amount of television* people watched during the immediate couple of days after the events. In other words, people in the middle of Nebraska watching extreme amounts of television tended to be at higher risk for posttraumatic stress disorder than many people who actually witnessed the World Trade Center collapse or the plane crash at the Pentagon.

Does this make sense?

One explanation is that people who were outside and witnessed buildings falling—but went home and didn't watch any television—experienced a horrible event, but they experienced it only once.

People who watched extreme amounts of television, however, "experienced" it hundreds of times. They also witnessed it with dramatic music playing in the background. And they may have seen as much or even more detail than eyewitnesses. TV viewers saw up-close images of the faces of people who were burning and bleeding. They repeatedly watched people jumping to their death. And they saw that same content over and over.

This study reminds us that, in today's media-driven world, the representation of an event can be just as powerful as the real thing.

When people tell stories about negative events, it's still scary, but we don't actually hear or see the event happen. We can only re-create the event in our imagination, and this was all people could do for most of human history.

Then came mail. In 1497, Emperor Maximilian I of the Holy Roman Empire established what was probably the first major postal service. Mail expanded greatly over the next several hundred years. You could get letters that chronicled harrowing events, family illnesses, and other tragedies. These were also powerful, but they didn't include audiovisual components, and mail came relatively infrequently.

Radio came into its own in the early 1900s. Broadcasts were infrequent and fuzzy, but they still could stir emotions. Perhaps the most famous example is the "War of the Worlds" broadcast on Sunday, October 30, 1938. For a special Halloween treat, Orson Welles and the Mercury Theatre created a brilliant spectacle that mimicked an ordinary broadcast being interrupted by fake "live bulletins," during which a frantic reporter described seeing a Martian invasion. This caused many people to panic and become truly worried for their lives, though the scale of the panic has since been debated.

Then, in the 1950s, television added visuals, and this brought the ability to frighten to new heights. Even in the early days, when televisions were small and black-and-white, and the signal was fuzzy, TV news producers knew how to use fear as a motivator. The TV news emphasized tragic events to increase viewership, and this resulted in dramatic anxieties around the economy, the Russians, and much more. Of course, news programs discussed real events, but television shaped and curated that news to increase ratings and earn money.

These examples show how the progression of media technology advanced what can be called the "Mean World Syndrome." George Gerbner, a communications scholar who studied the cumulative influence of television violence on viewers, coined this term in the 1970s. He and his colleagues showed that fictional and overrepresented news violence can significantly change someone's view of the whole world.

It makes sense. It's estimated that the average youth before turning eighteen will see about sixteen thousand murders on television, even if they never see a single one in person. These exposures are influential in shaping that young person's worldview. The world certainly does contain violence. But Mean World Syndrome can result in a person getting a *disproportionate* sense of the violence in the world. To the extreme, it can result in someone experiencing nightmares, desensitization to violence, and an irrational fear of being harmed.

Today's technological advances have amplified the potential effect of Mean World Syndrome. Two decades ago, the average television size was about twenty-four inches. Now it's double that size—around fifty inches—with more zoom and clarity than ever before, and you can even install surround sound to immerse yourself in the experience. With remarkable technology like drones capturing hard-to-reach footage and the ubiquity of cell phones, it's also easier to get longer, closer, and more graphic footage of events in real time.

All these factors contribute to a spectacle as real as memory, and at times even more influential.

Finally, and probably most important, media content is available twenty-four hours a day, seven days a week. This continuous barrage is what was so effective at catalyzing fear after 9/11.

This also affected people during the Covid-19 pandemic, especially during the first several months of 2020. Early in the crisis, new information was constantly unfolding, and that information was critical: It related to how people could protect themselves and their families from a deadly virus. But if an eight-ounce glass of information was sufficient to know what to do, many people were guzzling gallons of information all day and all night—and often from so many sources, some questionable, that people weren't sure what to believe.

The terms *doomscrolling* and *doomsurfing* refer to when people chug through huge swaths of negative news—even when they know that news is depressing, infuriating, and not helpful. This became a common experience during the worst times of the Covid-19 pandemic, and it will continue to be a danger during future tragic events.

The key to avoiding this is to find a *balance*. While there's nothing wrong with seeking out information and guidance, we need to proactively recognize and curtail doomscrolling when we recognize it in ourselves.

SOCIAL MEDIA UPS THE ANTE

Social media exists 24/7, and it features the same powerful cocktail of visual information, because traditional media content is embedded in social media. The screen isn't fifty inches on the diagonal, but it's five inches from our face, so it might as well be.

And we put this content in front of our faces *a lot*. The actual numbers change quickly, but currently the average time we spend on our phones overall is between four and five hours per day. And the average time we spend on social media is between two and three hours per day. For people between the ages of sixteen and twenty-four, the social media number is about an hour more. That is a lot of time. For many people, it's more time than they spend doing anything other than sleeping.

It's also worth noting that the numbers are increasing steadily. In 2012, the amount of time we spent on social media was about 90 minutes. In 2016, it was about 120 minutes. And in 2020, it was about 150 minutes.

Then there's the issue of frequency (which is the focus of chapter 10). According to a 2019 study conducted by global tech company Asurion, we check our phones an average of ninety-six times a day, about once every ten minutes. So, it's not just that we're spending a vast amount of time in total. It's also that we're basically never mentally free from our phones.

All this is possible because of today's digital technology. More than ever before, media is clearer, more compelling, more accessible, and larger than life.

When it comes to social media, there's also the special power that comes with how *personal* it can be. When I'm watching a tragedy on television, I may not have a direct personal connection to it. The tragedy may affect me deeply, but I won't necessarily feel like it is happening directly to me.

When it comes to social media, I'm not just seeing a horrible event; I'm also hearing about it from a real person I know, care about, and identify with. I can also read the comments from other people I know about how terrible they feel. Now, I'm affected not only by the power of the tragedy itself—but also by the effect it has on other people I care about.

All of this makes the experience feel more intimate, personal, and real.

This is why we need to remain aware of how dramatically social media can affect us. That doesn't mean we shouldn't use social media to share difficult stories, negative events, anger, and grief. As I discuss in Chapter 38, social media can provide critical comfort when we are struggling. Nevertheless, we must protect ourselves against both Mean World Syndrome and social comparisons. We must learn to buffer the risk to our own mental and physical health while using social media.

6
The Good Stuff

IF SEEING A LOT OF *positive* messages on social media is bad for us, and seeing a lot of *negative* messages on social media is bad for us, what good could we possibly get out of using social media?

Actually, a lot. As researchers, my team and I wanted to hear from real people about how they used social media in *positive* ways, so we conducted a large qualitative study that surveyed a couple thousand people from all over the United States. Each person was asked to anonymously express whatever came to mind about the relationship between social media and their mental health. In this chapter, I share the ten *positive* themes people expressed over and over.

1: "Social media helps me stay connected to family and friends who don't live nearby"

In today's world, we move around a lot. As of 2020, people moved an average of about twelve times in their lifetime. We no longer live in a world where people typically stay at a job for decades or live in one place all their lives. More than ever before, we are physically distant from many people we know.

In the past, when technologies didn't allow for effortless communication, old friends often became lost. Now, social media has made it easier to keep up with older and longer-standing relationships. This is something people in our study deeply valued.

Friendships formed early in life can be particularly powerful. Those friends often know us *well.* They have known us across time and can understand us as a more complete person. They know our successes, but they also know our challenges, flaws, and formative experiences. Digital technology offers a useful window into the past that we didn't have before cell phones, social media, and the internet.

2: "Social media helps me deal with the deaths of loved ones"

Social media doesn't immediately seem related to death. But consider the fact that, by about 2065, there are expected to be more Facebook accounts of dead users than living ones.

My colleague Beth Hoffman and I recently published a paper called "Their Page Is Still Up: Social Media and Coping with the Loss of a Loved One." We found that 37 percent of young adults recently experienced the death of a social media contact who was also a loved one. For some of these people, social media *complicated* the grieving process. For example, they may have become upset when a social media platform announced—with balloons, confetti, and streamers—that today was the dead person's birthday.

However, many people also found that social media helped them *cope with* a death. Sometimes it helped disseminate information about a funeral or service for that person. Other times, old images on the page of the deceased brought comfort, and the presence of that page helped friends and family members continue to connect with one another and share memories.

It will be interesting to see how social media platforms deal with death in the future. In the meantime, for people who want it to, social media offers a way to preserve the memories of loved ones who have passed on.

3: "Social media connects me to my international friends"

We surveyed only US residents, but the topic of international friendships and connections came up frequently. A couple of decades ago, it was uncommon for US college students to leave the country during their studies. Now it's estimated that about one in six undergraduate students has at least one study-abroad experience.

It's not just college students. In just one year—from 2018 to 2019—the number of US citizens who traveled abroad increased by 6 percent, reaching a total of over ninety million people. During that period, every world region experienced similar, if not greater, growth in international travel. For obvious reasons, this trend paused during the 2020 Covid-19 pandemic, but international travel is poised to rebound robustly.

Social media is an impressive way for people to extend and enhance their international experiences—before, during, and after they travel. I taught in West Africa almost thirty years ago, yet I'm still in touch with people from that formative experience.

Finally, don't forget that about one in seven people living in the US is a recent immigrant. Many individuals in our study expressed how glad they were that they could use social media to keep in touch with friends and family in their native countries.

4: "I can meet others in my health situation, which helps me not feel alone"

There's an interesting paradox in medical science, and it relates to rare diseases.

There are, of course, many *common* diseases in the United States. One is type 2 diabetes, which about thirty million Americans have. Therefore, even though it's a challenging condition to have, there are frequent studies related to it and plenty of support groups.

On the other hand, there are *rare diseases* like ribose-5-phosphate isomerase deficiency, which has been diagnosed only once in history. Progeria is another example of a rare condition; it affects about one in four million people. It's a terrible disorder that causes premature aging and results in death, usually in the teen years. Huntington's disease—which causes progressive mental and physical disability because of a breakdown of nerve cells—is also classified as rare because it affects only about one in twenty thousand people.

If conditions like these are so rare, why should we be concerned about them? Shouldn't we focus instead on things like type 2 diabetes, which affects thirty million people?

The interesting thing is that, *taken as a whole*, guess how many people are affected by what are considered rare diseases? About thirty million people—the same number of people who have type 2 diabetes.

This is where social media steps in. It can connect people who might not have enough support in their immediate vicinity. Before social media, people with rare conditions couldn't communicate regularly with others to discuss their shared conditions and motivate one another. They didn't have the ability to find out about research studies or new treatments focused on their conditions.

But now, for example, there is an active Facebook group for the National Organization for Rare Disorders. This group enables people to share information about and successes with genetic therapy for conditions like those mentioned above. People can access professionals who are interested in treating and conducting research on rare disorders. They can start fundraisers for specific conditions. They can also find others who are suffering with the same condition.

This illustrates that no matter how few people are dealing with something—or how unique we feel our needs are—social media can help link us to resources and support.

5: "Social media motivates me"

Chapter 4 discusses how overexposure to idealized portrayals of others can make us feel inadequate—and how this can exacerbate or incite depression or anxiety.

But there is a flip side to this.

Being exposed to others' achievements can also be *motivating*. How many Pinterest posts have inspired people to learn to sew a quilt from their old T-shirts? How many posts from friends reenrolling in school have helped people overcome their anxiety and take courses to fulfill their dreams? How many fitness success stories posted online have prompted others to try out Pilates or yoga?

Many of the people in our study used social media for encouragement. It is possible to let others' successes *drive us instead of deflate us*. We can appreciate others' accomplishments as a sign that we, too, can attain similar success. This uses comparison in an empowering way instead of a demoralizing one.

6: "Social media tells me about events"

Many people in our study discussed how social media is often the first place they learn about events in their local communities. That's because many events these days are only promoted via social media. Even high-level publicity often uses the strategy of creating a social media–based event and then disseminating that widely on various other platforms—whether they're promoting a major arena's concert or a local poetry reading. We also found that many of the people in our study planned and implemented their own gatherings and parties using social media.

Now that the Covid-19 pandemic has made video conferencing ubiquitous, a new world has opened up with virtual and remote events. Maybe you greatly enjoy monthly Scrabble tournaments at your local library, but during the pandemic, these were shut down. However, with social media, you can attend virtual events on the Internet Scrabble Club (isc.ro) and then share a screenshot of putting QUIXOTIC on a triple-word score with the dedicated Reddit Scrabble community.

Whether for virtual or in-person events, big-ticket shows or intimate gatherings, social media has become a major source of event information.

7: "Social media helps me find GoFundMe pages"

Social media has completely changed the face of fundraising and crowdsourcing. Take the example of the "ice bucket challenge."

This viral internet phenomenon brought unprecedented attention and funding to what for many people was a previously unknown condition—amyotrophic lateral sclerosis (ALS). Also called Lou Gehrig's disease, named after the great baseball player who died from ALS, it attacks the body's nerves and is universally fatal.

The origins and development of the ALS ice bucket challenge are somewhat unclear. In the spring of 2014, certain individuals donated to various charities after being doused with water. Some of these episodes were shared on social media.

The challenge then entered the mainstream when television anchor Matt Lauer performed it on the *Today Show* on July 15, 2014. At that time, however, the challenge was not associated with ALS. It was only later that day that golfer Chris Kennedy did the challenge and passed it on to his cousin, whose husband suffered with ALS.

From that point, the challenge became associated with ALS and resulted in record donations. There were over two million videos tagged with the ice bucket challenge on Facebook alone, and it raised over two hundred million dollars worldwide. This amount dwarfed what they had raised in previous years.

While this is an extreme example, it reminds us that social media is now a critical tool for helping us raise funds and support important causes.

8: "Social media helps me interact with others who share the same interests"

Participants in our study also praised social media for its ability to connect people with similar hobbies, passions, and ideas.

For example, on Reddit you can subscribe to specific "subreddits" about whatever you could possibly want to discuss and learn about. Some subreddits include "Coding," "Dungeons and Dragons,"

"3D Printing," "Lithuania," "80s Music," "Plants and Fungi," "Crappy Design," and "Shower Thoughts" (defined as "miniature epiphanies that make the mundane more interesting").

If you're a fan of lepidopterology—the study of butterflies—you might not find any fellow lepidopterologists living in your neighborhood, but one quick search on social media can put you in touch with other butterfly lovers quickly.

Just as you pick out what you wear every day, you can also select the content you surround yourself with online, so why not relate it to your passions?

9: "Social media raises awareness for causes"

The hashtag #BlackLivesMatter developed in 2013 after the shooting of seventeen-year-old Trayvon Martin. The hashtag rapidly spread through social media and inspired direct activism, including demonstrations, informational campaigns, and other politically influential actions. Nearly a decade later, the hashtag remains active and influential, in part because the deaths of George Floyd, Breonna Taylor, and Rayshard Brooks in the spring and summer of 2020 led to a remarkable wave of protests over racial justice and equality. This led to the removal of Confederate statues from prominent cities, policy changes, and much more. This is a key example of how the connected nature of social media can promote direct action in ways that previously would not have been possible.

Other examples of influential online movements include the #MeToo movement, Occupy Wall Street, the March for Science, and #LoveWins, all of which would not have been possible without the social media infrastructures and connections we have today.

It's worth noting that the power of these movements is directly related to how *organic* they are. In the past, companies have spent billions of dollars to generate buzz, convince people to believe something, or create a call to action. In 2018, for example, the tobacco industry spent over nine billion dollars in the United States alone to try to convince people that their deadly products are a good idea to buy and consume.

However, the spread of social media hashtags is largely governed by how much that idea resonates with people—not the sheer power of marketing dollars. For example, after rampant sexual abuse by Harvey Weinstein and others was revealed in October 2017, actress Alyssa Milano posted on Twitter: "If all the women who have been sexually harassed or assaulted wrote 'Me too' as a status, we might give people a sense of the magnitude of the problem." What followed was an outpouring that stemmed from people's lived experience. The following day, the phrase "Me too" was in over twelve million Facebook posts.

10: "Social media helps me get news"

Visiting the *New York Times* website doesn't really count as social media, but that's not how many people get their news these days. Instead, news is often spread through social media networks.

One reason people in our study said they prefer to get news from social media is that they like finding out about things exactly as they happen. A few decades ago, people often had to wait until the following morning (gasp!) to find out the results of baseball games, major trials, and financial data reports in the newspaper. Now, information is available immediately—not only from established news sources but also from whomever is in the right place at the right time with a camera phone. The flip side of this is that poor-quality or inaccurate reporting can easily be propagated by social media.

Study participants also liked that they could easily customize the *type* of news they get. It's easy to curate various social media sources—through subscriptions, shortcuts, and other means—to focus on news from a particular perspective or geographic area. This is particularly compelling to Americans, who like things customized.

Starbucks claims that they offer over eighty-seven thousand ways to customize their beverages—and they are just a coffee shop! In today's world people don't just customize traditional things—like the features of your computer or the toppings on your burger. They customize M&Ms with their own text and photos—and of course colors. You can buy a customized jigsaw puzzle of your neighborhood precisely centered at your home.

At one time, a handful of newspapers gave almost the same news to everyone. Now every single person can use social media to tailor their sources and get exactly the news they want.

It's important to remember that, while social media can negatively affect emotional health, it can also positively enhance our lives.

With social media, we can find out immediately about things that matter to us. We can easily share our own news and thoughts, which can be empowering. We can aid in the healing of others who have endured natural disasters or other traumas. We can get recommendations for nearly anything and learn about any topic. Perhaps these things seem self-evident, but amid all the significant concerns people have with social media, it's important to highlight these benefits so we can magnify and enhance them.

That's the purpose of this book. With the right kinds of strategies—and by reprogramming ourselves to create better conscious habits—maybe we can help the positive outweigh the negative.

7

Why Yo-Yo Tech Diets Don't Work

SOCIAL MEDIA AND DIGITAL TECHNOLOGY offer great promise and many benefits, but they also come with, or can lead to, serious dangers. One day, we can become obsessed with a new gadget or social media platform, and for a few weeks, we're so excited we can't seem to leave it alone.

Then we might become disenchanted. Perhaps we realize we're spending all our time on something that isn't that meaningful or helpful. Maybe we notice ourselves falling into bad media habits, and we vow to delete the app, quit the account, or just "not look" for a week. But then suddenly we're back at it shortly thereafter.

Dizzy yet?

Social media lends itself to these sorts of "yo-yo" behaviors.

A classic example of this, of course, is the yo-yo diet. Motivated to lose a few pounds using a new diet, we buy a new book, DVD, or app subscription. After one or two weeks, we're doing great. Then something interrupts our groove, and suddenly we remember how good pecan pie is—especially when it's consumed in whole-pie portions. And that's that. We overindulge, become discouraged, and lose our momentum. We return to old habits until we decide to go all-in on the next fad diet a few months later.

Whenever there are great positives or great negatives associated with a behavior, it's easy to fall into yo-yo situations, and it's the same with social media and digital technology.

However, something specific happens when we reduce social media use. After several weeks or months, we can start to feel like we're missing out—so we ramp up our use of the platform again. "Fear of missing out" (FOMO) is not a new phenomenon. It's been an issue in health research for a while because of how influential it can be in our decision-making processes.

For example, studies show that FOMO is an important driver of negative consequences of alcohol use among young people. Someone might attend a party they otherwise would not have because they think they might be "missing something" by staying home. The person is then more likely to drink because, well, everyone else is drinking. This drinking can then lead to some kind of negative consequence: getting sick, saying or doing something embarrassing, getting injured, or acquiring an increased risk of future alcohol abuse.

FOMO is also influential when it comes to social media. Hard-working, talented designers skillfully use various techniques to make platforms seem like parties. Colors, interactivity, logos, sounds, icons, and layout all combine with one purpose: to make the user want to stay. The platform is meant to be "sticky"—to stick a user there as long as possible. Designers want us to feel like closing that window or app will suddenly make us unpopular or out of the loop.

Beyond the compelling *format*, there's also the compelling *information*. Social media sites provide exactly the kind of irresistible news and information we crave. Humans by nature crave information, especially about other people we know. Hence the pervasiveness of gossip. As the "father of sociobiology" E. O. Wilson noted, people who gossiped had an evolutionary survival advantage. According to Wilson, gossip was "a runaway competition in who could be master of the art of social manipulation, relationship aggression, and reputation management." This certainly sounds like a good description of social media.

You can see how the yo-yo forms. We crave the information—because of both the design and the biological urge—so we relent and join the social media fray. Then we get disenchanted and swear it off. Then our biological and addictive urges, and FOMO, bring us back to social media, and the process starts all over again.

THE PROBLEMS WITH THE UP *AND* THE DOWN

Yo-yo dieting—the pattern of losing weight on an excessive diet and then regaining it when we fall off the wagon—doesn't work for sustained weight loss. This is well-known in humans, but it's also been observed in animal studies. When animals cycle through periods of scarcity and periods of bingeing, they end up gaining weight and developing other physical problems.

This is because problematic chemical changes happen in the body during both the loss *and* the gain. When people lose weight quickly, they often lose not only fat but also muscle, harming overall fitness and body functioning. Then comes the double whammy: The body's built-in compensation mechanism kicks in, and appetite skyrockets. This can lead to eating not only more overall but also more unhealthy stuff. This is why yo-yo dieters tend to have higher body-fat and belly-fat percentages. This style of dieting also increases the risk of things like diabetes and heart problems.

Mentally, yo-yo dieting can result in frustration and guilt. We think: *I lost ten pounds, but it came right back. Why can't I keep it off?* Someone might interpret this as a personal failure of character and not a consequence of things like blood chemicals. It also can make people feel like they're constantly focusing their mind on eating and nothing else. If someone gets into a consistent routine around food, they can focus on other aspects of their life. But if they're yo-yo dieting, they're often focused on food and maintaining their diet.

The same frustrations arise when we can't sustain a reasonable *social media* diet. Our yo-yo media dieting may set us up for endless cycles of disappointment one day and being hooked the next, never truly allowing us to be in a place of comfort. We're either awkwardly relearning lingo and idiosyncrasies of the platform—or we're trying to wrest ourselves from it.

Just like it is with food, social media also creates problems during both the up-and-down periods of the yo-yo cycle. Stopping use abruptly after having been a frequent user can be jarring. Someone might swear off all social media to improve their life, but they may not realize how much they came to depend on it. This can lead to cravings and anxiety—and then guilt and remorse when they give in and return.

Then, just like during the weight-gain phase of a yo-yo diet, rapidly returning to social media isn't usually done in a healthy way. When people quit a crash diet, they don't tend to return to a balanced, sustainable, high-protein, low-fat diet. Instead, they indulge whatever they've been missing out on. Just as a former dieter might start inhaling pints of Ben & Jerry's ice cream, someone quitting a social media fast might end up gorging problematic content for hours.

Clearly, we need balance. But is that even *possible* when it comes to social media?

8

The Social Media "Food Pyramid"

AS WE'VE DISCUSSED, FOOD MAKES a good analogy for social media. Like food, social media *can* nourish us. A balanced diet *can* be healthy. But as with food, it's easy to fall into bad patterns, like when we eat too much of the wrong things. This is why we need a "food pyramid" for social media.

The concept of the food pyramid was developed in 1992. The goal was to provide evidence-based nutritional guidelines in an easy-to-remember way.

The same standards apply for the social media pyramid I propose in this book. My hope is that this pyramid can be your key to curating your own personalized tech diet. However, it's important to remember that social media has only been around for a couple of decades, and technology changes quickly. So, we'll need to continually revisit how social media affects us and how we can respond in a positive way.

The need to keep improving in this way was highlighted for me on my first day in medical school in 1995. In his first address to our class, the dean of students said something very unexpected: "Half of what we're going to teach you over the next four years is going to be wrong, but as of right now, we just don't know which half."

This turned out to be true; a number of things I learned in medical school have since been reversed. I was taught to prescribe hormone-replacement therapy for women past menopause, but now this is known to increase the risk of strokes. I was taught that no drug could prevent HIV, but now we know a combination of medications is effective in preventing HIV for people in high-risk situations. And I was taught that something called a "coronavirus" was merely a minor and harmless cause of the common cold. We all know how that one turned out.

After a brief pause and some confused looks from his students, the dean went on to explain, "So, it's important to be a lifelong learner and adapt to new findings." That is, we need to learn what is known at this point in history, but we also need to expect that knowledge will change. While this book's social media pyramid reflects the evidence and information we currently possess, we can also tweak it based on what we learn in the future.

THE SOCIAL MEDIA PYRAMID

In our information-saturated world, models that are unnecessarily complex often don't stick. This is why I've divided the social media pyramid into three key principles that are easy to remember. When it comes to digital media use, *be selective*, *be positive*, and *be creative*.

Be Selective

Selectivity isn't about deprivation. It's about empowering ourselves to *actively* select specific digital and social media experiences that are the most valuable for us.

Questions to ask include: How *much* should I be using social media? How *frequently* should I be using it? (Yes, these are different things.) Should I use many different platforms or focus on just a couple? Are there particular platforms I should stay away from? Are there times of day I should emphasize or avoid? Should I focus on a small group of friends or interact with a lot of people? Should I be more active or passive in my posting?

Selectivity makes up the solid base of the pyramid because it is critical in terms of forming our overall social media and digital diet—the raw materials. If we are what we click, then we should be selective about it.

Be Positive

Positivity represents how we *build on* that foundation—how we craft specific ways of interacting that will ultimately maximize the value of our digital lives.

Studies show that it's good to be positive overall. In a major study of nearly a hundred thousand women, just being optimistic gave them less likelihood of heart attacks and death. As we'll see, though, positivity may be *even more important* in the digital milieu than it is in the offline world. Almost everyone has witnessed—or experienced firsthand—the severe consequences of negativity online. In part 3, I break down why this happens, what it means for you, and what specific steps you can take to stay positive while still being true to your feelings and beliefs.

Be Creative

Creativity brings everything to the next level; it's the tip of the pyramid. How do you tailor your experiences to your specific personality? How do you easily find ways of discovering those less-known corners of digital life that may be the most valuable for you? How do you exhibit innovation in how you use platforms and tools so that they work for *you* and not for their marketers?

TAILORING SOCIAL MEDIA TO YOU

But here's the problem with a simple, generic pyramid: It's not *tailored to you*. Many prescriptive systems tend to offer very specific guidelines. For example, a nutrition plan might extol the virtues of whole grains, fish, nuts, and fresh fruits and vegetables.

But what if you can't eat gluten? What if you are vegetarian or vegan, but the system tells you to eat fish and chicken? What if you're allergic to apples, nuts, or dairy? A food pyramid offers general guidance, but you must curate your diet so that it's appropriate for you.

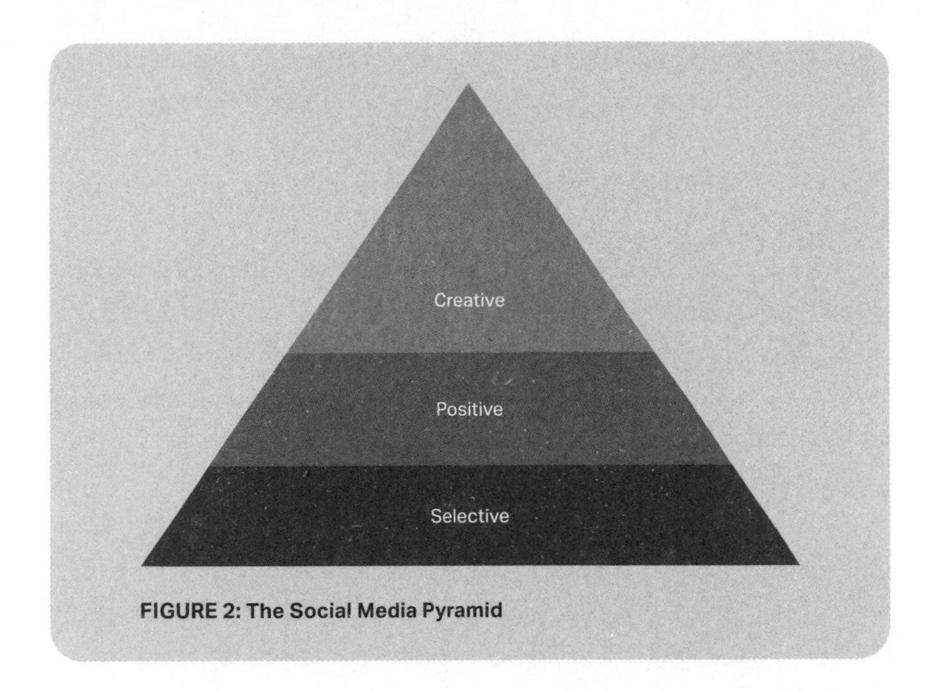

FIGURE 2: The Social Media Pyramid

Your digital diet is the same. My colleagues and I have found that people with different personalities are affected by social media in different ways, which makes sense if you think about it. If someone tends to be a worrier, exposure to endless threads focused on traumatic events will probably bother them more than others. As a result, in this book, I shy away from offering *very* specific or universal prescriptions. Instead, I encourage you to tailor your use to your own unique needs and situation—which applies not only to your personality but also to your occupation, schedule, aspirations, and more.

I understand the desire to quantify things. After all, we are limited creatures, and specific numbers can be practical and reassuring. But the issues and options in the digital media world are too complex. It's not just about the time you spend online; it's also about the frequency, the number of platforms, and the specific characteristics of those platforms. It's not just about *who* you interact with but *how* you interact with them, *when* you interact with them, and how you *balance* interacting with them online and offline. It's not just about what you communicate and how you communicate it. It's also about how your natural personality-based tendencies lead you to *interpret* what you experience. This is why this book will help you tailor the social media pyramid's three overall goals—being selective, positive, and creative—*to you.*

A final note: It's impossible to write a book about social media and not give examples that relate to specific platforms like Facebook, Instagram, Snapchat, Reddit, and TikTok. *But things change very quickly*. Consider the case of MySpace, which was the largest social networking site in the world between 2005 and 2008, but is virtually nonexistent now.

This is the reason I developed the principles in this book to be highly flexible across digital platforms and technologies. What works for how you interact with social media now might need to change in the future—as social media changes and as you reach different stages of your life.

It's time we learned how to take control of our relationship with technology, and the first step is to *be selective.*

Part 2

Be Selective

BEING SELECTIVE IS VERY DIFFERENT than cutting things out.

In behavioral science, we need to be careful about any instructions that focus on what *not* to do. Part of this is the human tendency toward *reactance*. The basic idea is that when we're told not to do something, we become tempted to do just that thing.

There is another human tendency that makes concentrating on what *not* to do somewhat dangerous:

when we are *reminded* of something, it takes center stage in our sphere of attention. Every time we tell ourselves not to do something, we may be unintentionally driving our focus to it.

Here is a good illustration. When I was in graduate school for education, one of my professors had worked closely with the producers of *Sesame Street.* One day, the education team had a brilliant idea. They decided to write a cute song about Ernie in the bathtub with his rubber duck demonstrating that you can't accidentally be sucked down the drain.

The problem was that kids, to that point, weren't concerned about whether they could go down the drain.

You can imagine what happened next. The makers of *Sesame Street* were soon flooded by phone calls from frantic parents stating that this episode had caused their children to forever be afraid that they might accidentally be sucked down the drain. Children all over the United States were refusing to bathe. Instead of quelling a fear, the episode had planted one.

In public health and behavioral science, telling people what *not* to do often backfires. For this reason, the first level of the pyramid is "be selective," that is, *proactively* make good choices.

This means: Focus on *actively picking* the social media you *do* want rather than focusing on what to *avoid*. This is especially important in today's society, which offers no shortage of information and entertainment. A USB drive the size of the end of my pinky finger can store enough material to keep someone occupied for several lifetimes.

What is limited, though, is our *attention*. So, be selective about where you spend that attention. You have the power to choose, so take charge of choosing. Don't worry about what you *can't* or *shouldn't* do; focus on what makes the best use of this precious personal resource.

Obviously, this applies to more than creating a balanced social media diet. This method can also be used to manage your use of another technology or another aspect of your life. Part 2 helps you curate your social media experience in a number of ways—optimizing, for example, when, where, and how much you use social media, as well as who you use it with. This sets the foundation for learning how to get more value and creativity out of your experiences on social media.

9
Select Your Time

IN INDUSTRY LINGO, WHEN SOCIAL media sites are said to be "sticky," what is supposed to stick to them is *you*.

After all, most platforms make more money for each additional minute you spend on their sites. Your attention drives up advertising revenue and share prices, which makes the company successful.

The only one who loses is *you*. If you get "stuck," you may miss out on more personally meaningful experiences.

Social media platforms are not inherently bad. We *can* learn to integrate them into our lives in healthy ways. However, we need to know what we are up against. Companies hire armies of talented designers, psychologists, and marketing experts to make their platforms highly engrossing. This is why most platforms can be scrolled endlessly.

The never-ending scroll feature didn't always exist. Most early platforms gave one page of results at a time—maybe ten or twenty results. Then, marketers learned that having to actively click to get to the next page represented an important decision node. A lot of people would stop looking between pages one and two, or between pages two and three. So why not make the first page endless?

This is similar to the autoplay feature of video streaming services. Once you're done with an episode, the designers know that if the next episode automatically starts while you're still scrambling for the remote, you're more likely to stay put and watch just one more. I recently saw someone wearing a T-shirt with red letters in the Netflix font on a black background that said "Nextfix." At times, this misspelling seems dangerously accurate.

Interactions on social media pages have become extremely sophisticated. For example, Facebook and Instagram notify users in real time when a response to a post, a new comment, or a new "like" appears. And that's just for starters.

You might get notified not only in response to your post but also in response to a comment you made on *someone else's* post. Or if someone posted in one of your groups. Or if someone you know "likes" a photo—even if it's not yours. Or if someone indicates that they are going to an event you said you *might* go to. And so on.

Of course, you can opt out of notifications. Easy, right? Not exactly. It can take as many as ten clicks and scrolls to find the right boxes. The complexity involved in turning these off—in addition to the fact that they are turned on in the first place by default—is part of the "sticky" strategy.

The sophistication doesn't end there. The next time you see a notification appear in real time, look at it carefully. Did it just appear, or did it fade in? How quickly did it fade in? Choices like this are proprietary. It's unlikely that companies like Facebook will be publishing a manifesto on how they determine the exact rate at which notifications fade in. Similarly, they won't be telling us how they choose the sounds that accompany their notifications.

However, social media companies probably think long and hard about these choices. After all, those notifications are critical. Over a billion go out every day, and each one is an opportunity to extend the amount of time that someone engages with the site. So, it makes sense to assume that the fade rate has been determined precisely. If it's too fast, it might appear jarring, which could result in people going to more effort to turn off notifications. But if it's too slow, it might feel awkward. Companies carefully consider and hone these decisions, spending as much as is needed to maximize how long people stay on their site. When you're running a multibillion-dollar business, you can't leave anything to chance.

This goes for the sounds that accompany notifications as well. Every sound has a precise pitch (how high or low), intensity (how loud), timbre (musical quality), attack rate (how quickly it starts), decay rate (how quickly it fades), sustain (how long it lasts), and more. These parameters are selected carefully to make you more likely to respond to it. Again, it's reasonable to assume that they don't just pick a premade sound from a digital library and go with that. They likely treat it more like developing a masterpiece.

The concept of stickiness did not begin with social media. While watching a television program, advertisements for the subsequent program mathematically increase the closer we get to the next program. If we're watching the 4 p.m. football game on Sunday and that game is set to end at 7 p.m., starting at about 6 p.m. advertisements for the next program will increasingly appear. That way, before football is over, we get more and more interested in whatever the next program has to offer.

For all these reasons, it can be difficult to stop watching, scrolling, or clicking. This is why one of the first things experts like to suggest is how much *time* we should aim for on social media. But what *is* the right amount of time?

In many ways, the optimal amount of time you should spend online isn't an exact number—it depends on what is specifically best for you. In fact, as we'll discuss, other aspects of social media use—like what you *do* on social media and *whom* you're doing it with—are probably more important than the precise amount of time you're spending.

Another reason I tend to avoid suggesting a specific number of minutes is that we didn't find a "sweet spot" in our studies. As I discuss in chapter 2, there seems to be a linear relationship between increased social media use and an increased risk for things like depression and loneliness (see figure 1, page 19). There is no ideal Goldilocks spot. Because of this, perhaps the best suggestion for how much time to spend on social media is "as little as possible," but that isn't very realistic or satisfying.

So, let's try to arrive at a ballpark number as a general guideline. The longitudinal study my research team and I conducted showed that people who used social media more than 300 minutes per day were about three times more likely to become depressed compared with those who used between zero and 120 minutes per day.

This suggests that a range of zero to 120 minutes per day might be preferred. To get even more specific, I'd suggest the middle of that range, or about an hour or so. That being said, more research definitely needs to be done to hone these kinds of recommendations.

About one hour each day. Is this feasible? The average use among young adults is more than triple that, so for some it may be a tough target to shoot for. It's important to keep in mind, though, that specific goals like this are rarely black-and-white. It's not like you're guaranteed to be fine if you never surpass sixty minutes of use—and you're doomed if you log sixty-one. Whatever feels healthy and balanced is the right amount for you.

Then again, the amount of time isn't all we need to be selective about.

10 Select Your Frequency

THE TOTAL AMOUNT OF TIME aside, a good question to consider is: How should you *divide up* that time during the day? In fact, it makes a big difference whether you lump the whole hour of social media into a single session or log on twenty times a day in three-minute bursts.

In addition to measuring time spent using social media, my research group also measured frequency. And no matter how many different ways we measured it, we nearly always found that increased frequency of social media use was related to an increase in anxiety, depression, and other poor emotional health outcomes. We also found that frequency often had an even *stronger relationship* with things like anxiety and depression than the amount of time people spent online. Why might this be?

One reason is our brains.

In *The Distracted Mind*, neuroscientist Adam Gazzaley and psychologist Larry Rosen explore how our tech-saturated world affects our brains. They show how the tasks of today overwhelm our brains, which evolved over millions of years to deal with a completely different external world.

For example, we evolved to take in a landscape of colors natural to the Earth—greens, blues, and browns—that fade to black each night at sundown. But today, marketers surround us with stimulating reds and yellows. Although there are reds and yellows in nature, they are relatively uncommon. There are red flowers, for instance, but they usually grow within a sea of green bushes, trees, and grasses. An orange sunset lasts only a few minutes, while the calming blue sky lasts hours.

And today we can completely ignore nighttime if we want, surrounding ourselves instead with artificial light and constant stimulation.

This stimulation confuses our brains. For example, the part of the brain responsible for sleep is called the pineal gland. The environmental signals it gets at sundown make it release a chemical called melatonin that helps us drift off to sleep. But if we don't experience sundown—if at midnight our room is as bright as noon and wild car chases are screeching across the television—the pineal gland isn't sure what to do. Our ancient brains simply aren't up to the new challenges posed by 24/7 tech.

Gazzaley and Rosen also provide a clue as to why frequency can be more influential than the amount of time when it comes to social media: The constant *shifting* of focus is particularly challenging for our brains.

Before screens, there was only one view: whatever was in front of us. Today, we shift rapidly among a half-dozen open computer windows. We have multiple apps active on our phones at once and continually toggle among them all. This constant shifting of attention takes a toll on our brains and can interrupt their normal functioning. Researchers still aren't clear on exactly how it happens, but this abnormal functioning can lead to things like depression and anxiety. Experts overwhelmingly agree that, while scientists are still learning more about these mechanisms, we should minimize multitasking as much as possible.

Yet social media use is defined by constant shifts in attention, which can interfere with everything else we're trying to do—whether that's performing well at work or school, achieving important goals, or maintaining personal relationships. Difficulties in these areas can also encourage depressive or anxious thoughts.

Another reason that frequent use might affect us is because frequent social media interruptions can get in the way of our brains *consolidating* important information. Our brains need time to "chew on" information in order to properly process that information and form memories.

Anyone who has played a musical instrument knows this from experience. Sometimes you get frustrated while practicing a difficult piece. But if you stop, take a break for a couple of hours, and then practice again, it's suddenly easier. It's like your brain has been practicing for you.

This isn't far from the truth. While the process of encoding memories continues to be mysterious, we do know the basics. First, new information enters short-term memory. This happens immediately when we experience something, and it involves many different parts of the brain, some of which have particularly fun names like the hippocampus and the amygdala. These short-term memories then need to be consolidated into long-term memory. This involves a complex rewiring that makes those long-term memories easier to retrieve later. Then short-term memory is freed up for tomorrow.

If this consolidation process is interrupted, say by constantly checking social media countless times throughout the day, both aspects of memory can suffer. Our short-term storage might be at capacity without as much room as we'd like, and the important stuff we want to remember might not make it into long-term memory. This leaves us with less mental energy and space to create, solve problems, and accomplish our goals. This is why it's important to give yourself uninterrupted consolidation time—both during the day and at night.

One of the times consolidation happens is during sleep. Experiments show that people remember different things in different ways before and after they've slept. Neuroscientists find that certain parts of the brain light up while we're sleeping, and that corresponds to how much we consolidate into long-term memory. In other words, sleep is essential for proper organization of memory.

It's also becoming clear that quiet time during the day is essential for *preparing* that sleep-time consolidation. In 2014, a group of Harvard neuroscientists published a study in the journal *Neuroimage*. Using specialized brain scans, they demonstrated that "resting state brain activity" was necessary to *prime* the brain for later consolidation of learning.

This means that taking breaks—those bits of downtime between our various daytime tasks, whether to visit the watercooler, go to the restroom, or take a walk—may help us process everything in order to ultimately make it stick. Some peaceful time is essential for processing all the other things we do.

This may be why people say, "I get my best ideas in the shower." Even when we aren't consciously trying to make breakthroughs, our minds are consolidating important things we learned—and this process can bring clarity to a situation and free up mental space for more creativity.

Frequent social media use is quite good at messing up these healthy processes. If we spend our precious break constantly checking our feed, we're not giving ourselves that important subconscious time to let those short-term memories seep into long-term storage.

When social media use is too frequent, we're also not giving ourselves time to *consciously* think about our lives, evaluate our actions, and make plans: *Should I follow up on that meeting? Should I text my friend to make sure he's okay? I get so anxious in that class; I'd better plan for how to study for the test.* Without time for self-talk like this, we can get overwhelmed, anxious, and depressed more easily.

All of this suggests that reducing the frequency of social media use may be better for our health and our overall happiness.

A good place to start may be to reserve social media use for only one to three sessions per day. This is consistent with advice from work productivity experts. For example, in *The 4-Hour Workweek,* Tim Ferriss advocates for checking email only twice a day so that it does not distract from more important matters.

Then again, if you're a social media marketer, and monitoring social media is part of your job, this isn't realistic. But that might make it even more important to ensure you have plenty of tech-free breaks during the day.

Nor is every day 100 percent the same. Some days or weeks might involve heavier social media use for any number of reasons—because of important national news, someone's birthday, or the holidays.

Still, try focusing social media into just a few sessions a day, and see if it helps unclutter your brain.

11

Select Your Platforms

IMAGINE TWO YOUNG MEN WHO each use social media about two hours per day. They are about the same age, and they do the same kinds of things on social media. In fact, the only difference between them is the *number of platforms* they use. One man divides his time between two platforms, while the other spends the same amount of time visiting eight platforms. Might this one difference in their social media habits—the number of platforms they use—lead one to feel more depressed, anxious, and lonely than the other?

Given what we've discussed so far, we might assume that using just two platforms would be healthier. After all, each platform represents its own little world. By spending more focused attention learning about these worlds, they might feel more manageable and familiar. We'd feel more "at home," know the written and unwritten rules better, and be less likely to make some sort of social gaffe.

Then again, might using eight different platforms offer its own benefits? For one thing, it would present more opportunities to find support and camaraderie, which could be healthier and more empowering. By using more platforms, we also might feel more tuned in to what is happening in the world, helping us feel more informed and in control.

To answer this question, my research team and I divided participants into four equal-sized groups according to the number of different platforms they used. The first group used no more than two platforms during an average week. The next groups used three to four platforms, then five to six, then seven to eleven. We only measured people's use of the top eleven most-popular platforms, which we knew accounted for more than 95 percent of all social media use.

Then we looked at the relationship between the number of platforms people used and their risk of depression and anxiety, all while adjusting statistically for everyone's time spent on social media.

The results were stark. It was another straight line—the more platforms people used, the higher their risk was. People who used seven or more platforms—compared with those who used just a couple of platforms—were about *three times as likely* to be depressed or anxious.

These results suggest that using fewer platforms is probably better for our mental health. This supports the first argument—that it's better to focus on just a couple of idiosyncratic worlds instead of spreading too thin among a bunch of them. When we use many different platforms, all of which have their own specific buttons and functions, we may be dividing our attention and energy in a problematic way.

Of course, there will be exceptions. In big studies like this, what we find are *tendencies*—people who use more platforms *tend* to be more depressed—but certainly some people use a lot of platforms and *don't* feel depressed.

Of course, how many platforms we use can depend on more than personal preference. Our boss might want us to have a strong presence on LinkedIn. We might use Twitter or Reddit to get national news but Instagram or Facebook for information about acquaintances. Our family might use WhatsApp while our best friends are on Snapchat. In today's world, we are encouraged to "collect" platforms, and some of that collection process might be unavoidable. But ideally, at least our choices can be *intentional*.

HONE YOUR PLATFORM COLLECTION

We tend to be reactive rather than proactive with what we do online. Have you ever joined a newsletter you never read, and yet you never unsubscribed? Do you scroll through a platform because it is easy to access and because you are automatically logged in—not because it actually brings value to your life? Do you find yourself sitting down at a computer for a particular reason, but then end up going down a rabbit hole because of autofill or "next for you" suggestions?

In the documentary *The Social Dilemma,* former insiders at the world's largest and most powerful digital technology companies describe in grim detail the lengths that their companies take to derail users from their actual intentions. They chronicle how designers routinely use algorithms specially designed to lure users into "rabbit holes" that will keep them occupied for the maximum time.

This is one of the reasons why we collect platforms. Platforms make it supremely easy to stroll along well-worn paths by removing every obstacle. I sometimes ponder why it takes me a dozen or more clicks and scrolls to use a piece of software at work to approve someone's purchase of an educational resource, but it only takes me one click to log on to and become immersed within a social media platform.

One way we can be more selective and intentional with our platforms is simply to be *fully aware of this process.* If we are cognizant that companies are actively trying to lure us into spending countless hours of our day on their platforms, we may be less likely to simply go along with their plans. After all, we will always have more meaningful goals and activities we can fill those countless hours with.

The next step in being selective about what you consume is to carefully consider which platforms stress you out more than they benefit you—and which ones are worth your attention.

Virtual-reality pioneer Jaron Lanier wrote a book called *Ten Arguments for Deleting Your Social Media Accounts Right Now.* After reading about his deep concerns related to the effect of social media on our personal lives, our political system, and even our sense of "truth," you just may decide to take his advice. However, we can also take a more nuanced approach—taking a scalpel to our digital media life instead of a battle axe—by trying the following exercise.

Start by writing down all the platforms that you use regularly. By regularly, I mean logging on and interacting at least a couple of times a week. I wouldn't count a *static profile*—for example, basic information about yourself on LinkedIn that functions as a professional resume. Rather, focus on the platforms that frequently take up your mental bandwidth.

Then do an experiment. Focus on using one of those platforms for a week and note two things: the *benefits* and *drawbacks* you perceived during and after using it. Did your experience on that platform motivate you or depress you? For instance, if you visited Pinterest regularly, did it inspire you to create something meaningful, or did it just make you feel bad that you don't have enough time for crafting?

If it's the former, Pinterest may be a platform that adds value to your life. If it's the latter, perhaps quit Pinterest and use that time to actually craft. Sometimes platforms can be so alluring that we spend more time mentally planning the things we want to do instead of actually doing those things.

When you do this, give yourself *a whole week at a time* for each platform. This is because your experience on one day might not be representative of your overall use. You could have a great day with one platform, while longer exposure might reveal problematic patterns.

After going through each platform, synthesize the information you now have. It's not as simple as giving each platform a thumbs-up or thumbs-down. Instead, put each platform in one of three categories:

1. **Cut:** These are the ones that you just don't need and would be happy to get rid of. Congratulations!

2. **Static:** These are the ones that you want to keep, even if you don't use them often. It may be a static profile on a LinkedIn page so

people can see where you work or a Facebook page where friends can see some pictures. You might update these once or twice a month, but probably not more than once a week.

3. **Active:** These are the ones you select to focus on for weekly to daily use.

Based on the study my team and I did, ideally you want to keep no more than two platforms in the active column at a time. That being said, we're all different and have different needs. Increase that number if this exercise reveals that more platforms truly contribute positively to your life.

Finally, once you've made your choices, try them out for a while and reevaluate. In isolation, each platform may have been beneficial. But when you combine them you might realize that one brings you more joy, value, and motivation than the others. Definitely select that one!

Selecting your platforms is a critical step because everything else you do will be based on those platforms. Given what you're up against—major social media companies that are financially motivated to reel you in—it can seem challenging at times. But by being more intentional about your selections, you'll be moving toward a more empowered and meaningful digital life.

12

Select Something Else Before Bed

WE HUMANS SLEEP FOR ABOUT A THIRD of our lives. Why?

The honest answer is that we don't really know. In 2005, the American Association for the Advancement of Science came up with a list of the top one hundred unanswered questions in science. "Why do we sleep?" made the list, along with things like "Is ours the only universe?" and "Is morality hardwired into the brain?"

We know that sleep "restores" us, but we don't really know *what* is being restored. We know that thousands of people die each year because of sleepiness—such as driving while drowsy—but we're not very good at helping people sleep better. We know that the average person has five or six REM cycles each night—in which your body is basically paralyzed as your eyes dart around—but we're not sure why this evolved in our species in the first place.

One thing we *do* know is that optimal sleep is crucial. Too little sleep is linked to heart problems, diabetes, stroke, obesity, depression, anxiety—and the list goes on. So, it's probably not surprising that too little sleep is also linked to dying early.

Interestingly, though, *too much* sleep is also linked to dying early. About a third of Americans get too little sleep, and about a third get too much. So, we're not strictly a sleep-deprived culture, as many people believe, but we are a culture that sleeps poorly. About two-thirds of us struggle with sleep at least once a week. This is a problem because the *quality* of sleep is also strongly linked to physical and emotional health. For example, compared with people who don't have insomnia, those with insomnia are about ten to twenty times as likely to have depression or anxiety.

Where does social media come in? Does it interfere with sleep?

It does seem to. First, as researchers have warned for years, looking at screens before bed can mess with the body's internal clock by reducing melatonin, a hormone that regulates sleep. Social media can keep us scrolling and mindlessly consuming media for hours, especially at night when we are tired and don't have the energy to stop ourselves.

But it's also possible that social media use *during the day*—even if it's not directly before sleep—can affect the brain enough to disrupt sleep.

My team and I have explored the relationship between social media use and disrupted sleep in a couple of key studies. In our first study, for which we were joined by sleep psychologist Jessica Levenson, we measured sleep disturbance among eighteen hundred young adults across the United States. We used a tool developed by the National Institutes of Health to look at things like sleep quality and whether sleep was refreshing. We found that 29 percent of our sample had what was considered high sleep disturbance. About 28 percent were in the medium sleep disturbance group, and 43 percent were in the low group.

Because we wanted to look at the relationship between sleep disturbance and social media use, we also measured everyone's amount and frequency of social media use.

People who used the most social media, compared with those who used the least social media, were about *twice* as likely to have more sleep problems.

But here's the interesting part. When we looked at the *frequency* of social media use throughout the day, the relationship was even stronger. Compared with the group who used social media at the lowest frequency, the highest-frequency group were about *three times* as likely to have poor sleep. So, the *frequency* of social media use seems to be even more closely linked to poor sleep than the time spent using social media.

How can daytime social media use affect our sleep, and why is frequency more of an issue than time?

We don't know all the specifics—sleep is just too poorly understood and social media is too new a phenomenon. We can, however, think about it in generalities. Chapter 10 discusses how our brains simply aren't up to the new challenges posed by frequently using technology and media that didn't exist until very recently. Those challenges may interrupt the complex and delicate brain conditions necessary for good sleep.

How can we optimize our social media use to maintain good sleep? To answer that question, we conducted another study. This time, as before, we measured social media use during the day and the quality of sleep, but we also measured how much social media people specifically used *during the last thirty minutes before bed.*

We discovered that—while the total amount of social media use over the course of a whole day was still linked to poor sleep—using social media in the thirty minutes before bed was *even more related* to poor sleep. The group of people who used the most social media throughout the day were 50 percent more likely to have poor sleep. But those who checked social media in the thirty minutes before bed were 62 percent more likely to have poor sleep.

Therefore, the last thirty minutes before bed seems to be particularly crucial. If our goal is to create a healthy, balanced plan for social media use, then one important way to protect the quality of our sleep is to avoid use in that half hour before bed.

THOSE PESKY THIRTY MINUTES

Why is it that those thirty minutes just before bed are so crucial to a restful night of sleep? There are a few possibilities. One is the whole "blue light" thing. As you may have heard, light that appears white to us actually has colors in it from all across the spectrum—from red at one end to blue at the other.

The light from our electronic devices comes from more of the blue side of the spectrum, as does light from the rising sun. The setting sun, however, produces light that's more on the red side, which is why the sunset often appears orange-ish.

Therefore, our species—and many others—evolved so that light on the red side of the spectrum signals our sleep centers to take over, and light on the blue side of the spectrum signals us to wake up.

One theory is that, by checking our phone before bed, we are accidentally activating brain centers that signal us to wake up.

In response, tech marketers have developed settings that produce a warmer red-type light in the evening. So far, however, there's little evidence that it actually helps. And skeptical scientists suggest that any light pollution right before bed isn't a good idea.

There's another important reason to avoid digital media at bedtime: If your plan is to sleep, then it's not a great time to get worked up or mentally stimulated about anything. And that's exactly what social media platforms are designed to be—highly stimulating—regardless of whether the light your device is emitting is redder or bluer.

The same issue applies, of course, to television, video games, and other electronic devices. About two-thirds of bedrooms have televisions—including kids' rooms—and many people, especially young boys, play video games for hours into the night.

What else is there to do during those thirty minutes before bed that will help you drift off to sleep without anxiety or concern?

First, read a book. And if you're worried that a mystery or thriller might be too stimulating, pick one that is calmer.

However, my number one suggestion is to keep a gratitude journal. Studies suggest that keeping a gratitude journal may help reduce the risk of heart attacks and depression, while improving sleep and even decreasing materialism. Many psychologists recommend it.

There's something about expressing gratitude in this way that helps us focus our brains on the positive things in our lives—a perfect way to end the day.

If scrolling social media right before bed has become a habit, it's time to develop a new one. For one week, try reading or journaling during that half hour before bed, and I'll bet you won't want to go back.

13

Select People You Know Well

IN TODAY'S WORLD, WE CAN be friends online with a lot of people we have never actually met face-to-face. In a historic context, this is extremely unusual. For most of human history, the only way to get to know someone was by seeing and hearing them directly.

With the invention of the telephone about a hundred years ago, we could talk to someone without seeing them. But we mostly talked with people we already knew. We didn't develop new relationships with people whom we talked with solely over the phone with no intention of ever meeting in person.

Today? Most people who use social media develop online-only friendships of one kind or another, and young people in particular are very comfortable connecting with many people they have never met and will never meet in person. People have built romantic relationships with—and even married—people they never actually met face-to-face.

Consider your contacts on social media. How many are people you have never met face-to-face? Maybe you liked a comment someone made, or you connected with a friend of a friend. Or perhaps you followed someone on Instagram who liked the same band you do, and then you felt that Instagram high when the person followed you back.

Does having many "non-face-to-face" friends—people that we've never actually met in real life—affect our emotional health? My colleague Ariel Shensa and I looked at this in a recently published study. We surveyed over a thousand young adults and asked them what percent of their friends they had *never* met face-to-face. For some people, the answer was zero. They strictly limited their social media circle to people they knew in person. For other people, though, that number was high. In fact, our participants reported having never met face-to-face an average of 39 percent of their social media contacts.

And there was a significant relationship between the number of non-face-to-face friends and mental health concerns. For every 10 percent increase in the proportion of friends a person did not know face-to-face, there was a 9 percent increase in the chances of depression.

One explanation for this is that, when we have never met a person face-to-face, the effect of social comparison—as discussed in chapter 3—can be exaggerated. When we know a person well, we know their faults and foibles as well as their best qualities. When that person posts a tremendously happy picture of their family, we know from personal experience that their family life is complex and not always rosy. When they post a flattering profile picture with not a hint of a double chin, we know that they don't always look this good, because we have *seen* that double chin in all its glory.

But with people we have never met before, we don't have a complete picture of their lives. We have no other context except their curated social media profiles. If we start comparing ourselves to them, when we know exactly how imperfect our own lives are, it will be easy to feel we don't measure up.

The relationship between having more non-face-to-face friends and depression also may be explained by *social network theory*. This theory says that "strong ties," which are based on trust and affection, offer excellent emotional support in uncertain situations. Most of the time, to develop these kinds of relationships, we need to experience people in person.

Social network theory doesn't think that "weak ties" are bad. In fact, they can be very helpful for finding new information and resources, like job leads and social activities. However, "weak ties" *don't provide the same support* in times of challenge. Friend networks with weak ties provide more superficial support and less substantial support, and there's a higher risk of miscommunication.

On the flip side, if more of our friends are considered "close," this may indeed translate into *improved* mental health. In that same study, we also asked participants what percent of their friends they consider "close," and the average response was 35 percent.

However, the really interesting thing was that the *higher* the percentage of friends someone considered close, the *less* likely they were to be depressed. Someone who said that 50 percent of their contacts were close tended to feel better in general than someone who said that only 30 percent of their friends were close.

This suggests that, to reduce our risk of problems, we should maximize the percentage of online connections we consider "close." In other words, it may be better to use social media to *extend* relationships that are already close than to use it to forge whole new relationships.

CAN YOU BE SELECTIVE AND STILL BE FRIENDS?

Limiting your social media circles to people you know well might be good advice, and it's easy to say, but how do we actually *do* this?

What you do depends on your situation. On the one hand, this advice isn't asking you to get rid of anyone you care about. There are many levels of friendship, and we *can* develop meaningful, genuine online-only friendships with people we've never met face-to-face. On the other hand, just because you follow someone on Instagram, and they follow you, doesn't mean you really know each other or are necessarily "friends." Connecting with someone once through social media doesn't mean you have to stay connected forever.

The goal is to be intentional and selective, so that you focus your social media use on people whom you like, make you feel good, and are genuine friends.

One option is to go through your social media, carefully select close followers and friends, and gently disconnect from the rest. This is like a form of decluttering. Author Marie Kondo instructs people to hold up every item of clothing and object in their house and ask, "Does this bring me joy?" If it doesn't, you should get rid of it. You can even use the same metric to decide who to keep in your social media circles, asking, "Does this person bring me joy?"

Having a honed list can translate into a more concentrated, positive, joyful experience online. Nevertheless, this process can cause anxiety and awkwardness if people find out they've been "unfriended" or "unfollowed." While social media platforms don't automatically alert people that they've been Marie Kondoed (yes, I'm using her name as a verb), people sometimes avoid winnowing their social media contacts for that reason.

Another option is to use your selected social media platforms to *focus*. Facebook and Instagram, for example, allow you to define "close friends." Some people make it a habit of sending routine timeline updates or pictures only to their "close friends" list. That way, they don't lose touch with their "weak ties," but they still *focus* on the strong ones.

You can also define groups for specific purposes. During 2020, after the Covid-19 pandemic led to at-home quarantines, I sent a message to my network asking who was interested in joining me in an exercise challenge to help us through that difficult time. I created a private group for only the people who responded directly. What ensued was a lot of fun, and using a private group was a great benefit for all. Everyone I contacted had asked specifically to be included, so I wasn't spamming anyone, and the people involved also felt free to interact with one another, since they knew that they all had similar interests.

Finally, one last suggestion is to do this regularly, like a spring cleaning. As you look through your lists and make active decisions about which connections are no longer serving you, you'll be increasing your "close friends" percentage and setting yourself up for improved mental health.

Remember, though, that creating a more intentional and selective list of friends, contacts, and groups is not only about throwing things out—it's also about remembering and appreciating what you *have*. Spring cleaning your lists can remind you of relationships you've neglected and want to deepen.

Yes, being selective about people can be difficult. Remember, though, that reconnecting through social media remains as easy as disconnecting. Just because you winnow your contact list in half doesn't mean that—like Thanos—you have finger-snapped those people away forever. If you miss someone and you want to get reacquainted, you can always reconnect. But for your own empowerment and emotional health, be selective with your lists.

14

Select Your Digital Holidays

THE WORD *HOLIDAY* COMES FROM "holy day." But what does "holy" mean in the first place? The primary definition in the twenty-first century is something like "dedicated to a religious purpose."

However, before the religious meaning became primary, Old English and German terms—like *hal* and *heil*—simply meant "whole" or "healthy." Not coincidentally, a holiday is a time to *recharge*. It is a time to rest up and regain your health. It is a time to become *whole* again instead of fragmented.

"Fragmented" is a somewhat accurate description of our lives in the twenty-first century. It's becoming harder to have a stretch of uninterrupted time to concentrate on any given thing. We have access to a huge list of contacts on our devices, and although this can be a good thing, it can also fragment our core relationships with family and our closest friends.

Even our own *identities* can feel fragmented. We have many different personas. We often find ourselves behaving differently at home, at work, and at social gatherings—becoming a particular version of ourselves to fit the situation. This continues on social media, with all the various ways we present ourselves on LinkedIn, Facebook, Twitter, Instagram, Snapchat, TikTok, and Reddit.

Keeping up with all these identities can be exhausting. We can't just relax into who we are at the core. In a sense, we're "playing a part" wherever we go—constantly having to remember what we said or did on this platform compared with another platform.

But social media "holidays" can help us become whole again. Maybe you've experienced this when it comes to work. A well-timed holiday can help you feel reenergized, so you go back to your routine a little less burned out. Similarly, a holiday from social media can reduce some of that "part-playing" and help you reground yourself in your true identity.

A social media holiday doesn't need to be a once-a-year trip to a quiet beach house. It can involve dedicated time off during the weekend to recharge and get ready for a new week. Holidays can also happen on a *daily* level. Most of us wake up and immediately dive into our days. But a good body of research shows that waking up early—in order to have some recharge time first thing—can reduce stress, procrastination, and negativity. In fact, no matter when it happens, any conscious recharge often helps us to be more effective, balanced, and agreeable during the day. Experts have suggested that this is how things like meditation, aerobic exercise, and yoga might function.

YEARLY, WEEKLY, DAILY

The positive effects of work and social media holidays are analogous. Just like breaks from our work routine are important at three levels—yearly, weekly, and daily—we should also consciously take a break from social media at each of these times.

Let's start with the annual level. Recent studies have shown that taking a sustained break from social media can result in significantly improved mood and outlook. In 2016, a Danish researcher conducted an experiment with over a thousand Facebook users. On average they were in their mid-thirties, had about 350 Facebook friends, and used Facebook for over an hour a day. They were randomly assigned to either take a weeklong break from Facebook or to keep using it like usual.

The results were dramatic. After the experiment, the people who took a break from Facebook had significantly better *overall life satisfaction*.

Another interesting finding of this study was that heavier Facebook users got the most benefit from taking a break. This might be like vacations from work: Compared with more balanced workers, true workaholics may get more benefit from taking some quality time off.

This study provides one model, but it's important to think carefully about your own situation. Is one week the right holiday for you, or should it be shorter or longer? Is once a year the right time frame for you, or should you take three three-day breaks four times a year? Or should it be a whole month once a year? This is something else to thoughtfully select.

If you do decide to take an extended social media holiday every year, you might want to do that in October. Why October? To show solidarity with the Offline October movement.

This movement began in 2017, when there were two high-profile tragic deaths linked to bullying on social media in Colorado. In response, friends of the deceased started a movement called "Offline October," during which they pledged not to use any social media. The program tagline is "Don't Post a Story, Live One." Individuals from over 150 schools representing most US states and eight countries around the world have pledged to continue to use October as their annual social media holiday.

Although there are fewer studies that speak to this specific issue, I think some sort of *weekly* break is valuable. It's not hard to see how weekly breaks from other routine activities dramatically affect our society. What would the week be like without a weekend? Why is it that major religious traditions include a weekly Sabbath, which comes from the ancient Hebrew for "rest" or "cessation"?

In 2014, I was honored to be included as a speaker at the TEDMED conference in San Francisco. One of the most compelling speakers I heard was Tiffany Shlain. A filmmaker by trade, she presented a captivating set of ideas about the brain's complexity, plasticity, and humanity. Shlain has developed her reputation as an internet leader and Renaissance woman by embracing technology. As a small example, she founded the Webby Awards, which celebrate the best web-based films each year, and she's won at least two dozen various awards for her films and documentaries. Yet even *she* completely unplugs from the internet for twenty-four hours each weekend.

This idea is echoed by J. Dana Trent, a Baptist minister who in 2017 wrote *For Sabbath's Sake: Embracing Your Need for Rest, Worship, and Community*. In this book, she argues for the shedding of social media once a week—also for twenty-four hours—in order to slow down. As a religious leader, Trent comes by her suggestion of a break every seven days honestly; seven is a well-known number in religious texts.

The seven-day week is also deeply rooted in the social media world. For example, many of the most common and popular Instagram and other hashtags are #MCM (Man Crush Mondays), #TT (Transformation Tuesdays), #WCW (Woman Crush Wednesdays), #TBT (Throwback Thursdays), and #TGIF (Thank God It's Friday). Although Instagram's parent company is not likely to promote it, #SMS (Social Media Sabbath) might be the healthiest hashtag of all.

Most importantly, you should *personalize* how this works within your world. One solution can be to piggyback a weekly social media holiday onto other weekly events. For people who practice a religion, that weekly event can be related to religious happenings. But it is just as easy to find other weekly, nonreligious rituals that lend themselves to some respite from social media.

This weekly break doesn't need to be for twenty-four hours, either. Try Sunday before dinner, Wednesday after work, or Friday night to Saturday afternoon. The point is to make a feasible plan that you can commit to rather than fulfilling an exact number of hours.

What about taking social media respites during each *day*? Again, options abound regarding when might be best to protect your time. Many people consciously avoid social media in the early morning, since this is already a time when many people recharge and prepare for their day. Research shows that we tend to be most productive in the morning, so this makes it a particularly good time to avoid social media distractions and stimulation.

Another good time might be late afternoon, when many people tend to have a "rebound" energy peak. Finally, taking social media breaks at night can help improve sleep (as chapter 12 discusses).

I encourage you to think of these holidays as *positive* experiences in which you are *selecting* something rather than rejecting something. When we take a break from work, whether it's a vacation, a weekend, or a relaxing evening, we don't think of that as restrictive.

Social media holidays should be similarly freeing, relaxing, and recharging. What you do is up to you. Walk in nature or ride a bike. Focus on a creative project, like art or music. Visit with people over coffee and have a conversation that doesn't require emoji to convey feelings.

And remember that you can always hone your holidays over time. Most people don't take the same vacation forever. You might need to adjust as you learn what works best for you.

However, ideally, choose digital holidays that fit all three various time periods—yearly, weekly, and daily—and fill that time with something you truly enjoy or that makes you excited. That way you won't think about what you're "missing," only what you're *adding* to your life that social media can't provide.

And don't worry, when each break is over, #MCM, #TT, #WCW, #TBT, and #TGIF will still be waiting for you.

Part 3

Be Positive

If you realized how powerful your thoughts were, you'd never think a negative thought again.

—*Peace Pilgrim*

POSITIVE PSYCHOLOGY IS HAVING ITS DAY in the sun—or more accurately, its couple of decades in the sun.

Being positive, it turns out, saves lives. Various studies suggest that optimism may increase and improve work productivity, the likelihood of financial success, strong friendships, social lives, and even life spans. Because of this, psychologists increasingly encourage people to foster positivity, optimism, and happiness.

This might seem like an obvious thing for psychologists to do, but it's actually revolutionary. In the past, psychology has focused mostly on "fixing" problematic ways of thinking. But positive psychology is different. It prioritizes building strengths over fixing weaknesses—and on enlarging good things instead of merely repairing bad things. It also aims to help *everyone* become happier instead of only focusing on people with diagnosable disorders.

This way of thinking has only come into the foreground over the past twenty years—after Martin Seligman and Mihaly Csikszentmihalyi published a seminal paper that revolutionized positive psychology.

Positive psychologists encourage things like keeping gratitude journals, which have helped relieve depression in studies. Positive psychologists also encourage acts of kindness and volunteer service, knowing that these kinds of activities tend to help the giver as much as the receiver. They also help people nurture the positive things they already do, so that those areas of their life can grow.

This doesn't mean that positive psychologists are unrealistic Pollyannas who think we should wear a forced smile all day. In fact, they discourage unrealistic optimism and instead encourage people to consciously acknowledge negative emotions.

In other words, they help people find and grow *natural and authentic* positivity. Martin Seligman came up with the acronym PERMA to summarize how to do this.

"P" stands for consciously seeking out and enjoying *positive emotions*.

"E" encourages being fully *engaged* or absorbed in things we enjoy.

"R" advocates developing positive and substantive *relationships*.

"M" reminds us to actively seek *meaning* in our lives.

"A" helps us work toward having and celebrating *accomplishments*, even if they are small.

Each aspect of the acronym is something we should consciously seek out, work toward, grow, and remember in order to develop genuine positivity.

Focusing on the positive is effective in our everyday lives, but it is particularly powerful when it comes to social media. This is why positivity is the middle chunk of the social media pyramid. We can apply the principles of PERMA to improve our digital lives—though we also need to be conscious that the opposite effect can occur if we are not intentional.

Positive emotions: We can have experiences on social media that produce very positive emotions—or very negative ones.

Engagement: We can enjoy being fully engaged in certain social media experiences—or we can let social media distract us so much that we aren't fully engaged or immersed in other important things.

Relationships: We can grow and extend very positive relationships on social media—or we can allow it to make our relationships less authentic and less genuine.

Meaning: We can find meaning on social media—or we can fall into a pit of commercialism and meaninglessness.

Accomplishments: We can celebrate accomplishments on social media—or we can let social media consume so much of our attention that we fail to achieve things that would give us real satisfaction.

Part 3 is dedicated to positivity. It discusses the most recent research around positive and negative experiences on social media—and what that means for you. It explores how to *avoid threats* to positivity and *enhance opportunities* for positivity. But first, before reading any further, consider for yourself: What are some positive and negative experiences you've had on social media recently, and how have those affected you?

15
The Power of Negativity

BEFORE INTRODUCING MORE POSITIVITY INTO our digital lives, we first need to face the negativity that exists. It's sort of like preparing for a big football game. If we're going to win, we've got to carefully understand our opponent. Only then can we optimally stand up to them.

It's no surprise or revelation to state that there's a lot of negativity on social media. In fact, negativity in the digital realm has spawned a whole new vocabulary. Twenty years ago, "trolling" was something you might imagine only happened while playing Dungeons & Dragons. You might have mistaken "FOMO" as the name of a household cleaner. And "flaming" was merely a good description for something that a car did after it exploded.

As you reflect on your own positive and negative experiences on social media, try to give an estimated percentage to each—how much of what you have experienced would you consider positive and how much negative? A lot of what we experience on social media is relatively neutral, so our positive and negative numbers usually don't add up to 100 percent. For instance, you might feel that 60 percent of your experiences online are neutral, so how would you divide up the remaining 40 percent?

To better understand the effects of negativity, my team and I did a study in which we surveyed over a thousand young adults and asked them this question quantifying the percentages of their positive and negative experiences online.

Interestingly, we found that the percentage of negativity each person experienced translated into real emotional health concerns. Specifically, for every 10 percent increase in negative experiences people had, they were *15 percent more likely to be depressed.*

The numbers add up. Say a boy in middle school begins having negative experiences about half of the time he's on social media. He's bullied online by some classmates, he sees posts of friends hanging out without him, or he simply feels inadequate as he compares himself to others on social media. Because of compounding percentages, suddenly his risk for depression is double or more what it would have been without those negative social media experiences.

This means that the amount of negativity we encounter on social media is critical. But we also wondered if positive encounters might balance out the negative.

To a degree, they do. In our study, we *did* find that an increase in positive experiences reduces the risk of depression. But here's the catch: There was only a *very small reduction*—about 5 percent—in depression for every 10 percent increase in positive experiences. Thus, an increase in negative experiences comes along with about triple the change in depression that an increase in positive experience has (or 15 percent compared to 5 percent).

To make things worse, in other studies we conducted, we found *no* significant benefit to positive experiences. For instance, in a different study, we found that people who had negative experiences on social media were more likely to experience loneliness, which isn't so surprising. What *did* surprise us was that, unlike the study focused on depression, positive experiences *did not* reduce loneliness or offset bad experiences.

Taken together, these kinds of studies suggest that *negative experiences online may be substantially more powerful than positive experiences.*

This is consistent with "negativity bias," a psychological concept that refers to how negative events are often experienced as more powerful than equivalently positive events. For instance, say a student is taking five classes. In four of them, they are getting As, but in one they are averaging a C. Even though it is only one class, that bad grade will probably take up more than one-fifth of the person's mental energy. They may focus on that one negative experience almost exclusively, becoming anxious and upset, while dismissing or ignoring the positive grades.

Negativity bias might be even more powerful in the online world. For example, imagine someone spends an hour online. For most of that hour they are happily clicking "like" on pictures of cute puppies, energetic TikTok dances, and creative selfies. They may even have an uncommonly good experience—such as hearing from an old friend they had lost touch with.

But during that same hour a social media comment is misinterpreted, and an angry comment is posted in response. That retort soon snowballs into more angry comments, with no support or acknowledgment from friends.

That negative experience, even if it lasted only a few minutes, might stick with and bother that person for the rest of the day, week, or month. It might make them less confident—not only on that platform but also in other situations, including offline. It also might make the person wonder why none of their friends came to their defense, even though their friends might not have even seen the difficult exchange. As a result, that one negative interaction might cause lasting damage to the person's mood, perspective, and self-image.

Remember, this can happen even when the *majority* of someone's social media experiences are otherwise positive. Many images of cute puppies—and even a message from a long-lost friend—can't always balance one bad comment. That is how negativity bias works online.

This can become a vicious cycle. If someone gets down enough, they may not take opportunities to meet up with friends or seek support from loved ones, leading to more time online, more challenging experiences, and so on.

LEARNING TO PLAY THE VIOLIN

When learning anything new, it takes time to do it well. For example, when someone first starts playing the violin—scratching out the first few notes—it ain't pretty. But they persist, and over time, by both reducing poor technique and accentuating artistry and skill, they can become capable of producing beautiful, even sublime music. This is the key to understanding our current situation with social media: *We are still learning how to use it as a society.*

Since social media is still relatively new, it perhaps isn't surprising that negative experiences outweigh positive ones at this point. However, we can learn to dramatically change the equation by *buffering the negatives* and *accentuating the positives*. That's what the rest of part 3 focuses on. We can learn how to insulate ourselves against difficult content while at the same time increasing positive experiences.

The first step is to accept that we'll inevitably encounter negativity online. Initially, social media may not provide all that we're hoping for until we learn to play the instrument better. But with practice and effort, and by making conscious decisions about when and how we consume social media, we can succeed.

16
Find Your Negativity Threshold

OUR "NEGATIVITY THRESHOLD" IS NOT something that we consciously think about frequently, but subconsciously we think about it all the time. For our purposes here, I'll define it as the amount of negative experiences we're willing to have or create in a given amount of time.

We are constantly making conscious and subconscious decisions about whether to avoid or instigate any number of potentially negative experiences. When making small talk with coworkers before a meeting begins, we have to decide whether or not to address some hot-button issue that someone may have joked about. If someone carelessly bumps into us while shopping, we have to decide whether to hold our tongues to avoid an unpleasant exchange. At home, if we think our kids are watching too much Netflix on a given weekend, we have to decide whether to make a big to-do and risk a confrontation.

Whatever we do, we can't avoid negativity entirely, and sometimes our choices make things more tense or unpleasant. Nevertheless, everyone makes both conscious and unconscious efforts to minimize negativity, even when we choose to create conflict.

Given this, why do our digital and social media worlds facilitate so much more negativity than the offline world?

Part of it is how easy it is to misinterpret what people say online, and part of it is simple statistics. Even the most positive and polite person can be misinterpreted in the social media world, and there's a strong mathematical probability that *everyone* will offend *someone* at some point online—even when we don't intend to.

Take, for example, the most common and non-offensive posts you can imagine. Like that old favorite: a photo of a meal. Or the classic warm-and-fuzzy: two pets cuddling. Or the universal tearjerker: a meaningful quotation about how love will triumph over anger.

In each case, despite our intentions, some people will be put off or hurt. If the meal includes a hamburger, vegans may find it distasteful. Cuddling dogs might remind someone of their lost pet. The inspiring quotation could rub someone the wrong way if, unbeknownst to us, they're going through a nasty breakup.

We can even put some numbers to the likelihood of inadvertently offending someone with our well-meaning, happy post.

The average Facebook user has about 350 friends. Let's assume that not all 350 will see something someone's posted. Maybe only about 20 percent, or about 70 people. Further, let's estimate that there is just a one-in-twenty chance that someone will be offended, or 5 percent. That means each person has a 95 percent chance that they will *not* be offended. Therefore, the chance that *everyone* will be not offended is 0.95 multiplied by itself seventy times, which is 0.03, or 3 percent.

This means the chance that *at least one person will be offended is 97 percent*—a virtual certainty.

In other words, no matter how nice and friendly we try to be, it's not a question of *if* we will upset someone when we post on social media—it's a question of *when* and *how many*. We can estimate that about three or four out of seventy people will be bothered in some way by almost every post.

These are conservative numbers based on averages. What if you have more friends? What if you also post on Twitter, where you have no idea how many people are seeing it? Then the possibility of rubbing more people the wrong way increases.

Most of the time, people keep their reactions to themselves, and offenses go unnoticed. The vegan *might* comment negatively on the burger, but it's more likely that they will keep their irritation to themselves. Is that better or worse? We certainly don't want to be told every time we cause distress, but we also might change what we post if we knew we were causing offense—yet we aren't likely to ever know who and how many people are bothered by our posts.

This is where our personal negativity threshold comes in and influences the decisions we make. For some people, the cost of offending *anyone* might be too great, and they might decide to quit all social media—their threshold for negativity is basically zero.

Others don't mind offending people, or they might even *want* to. For instance, some people know they hold unpopular beliefs and ideas, and they use social media as a tool for propagating and leveraging those beliefs. These people have a very high negativity threshold.

Most of us have a threshold somewhere in the middle. We can tolerate a certain amount of negativity, whether it's caused knowingly or inadvertently, but we pull back from situations we know will be very charged, posts that might offend a lot of people, or frays that could cause undue stress.

Every time you post on social media, the question you need to ask is: How willing are you to offend people? What is your negativity threshold? How willing are you to post something to brighten some people's day knowing it may be potentially irritating to others? This isn't a rhetorical question. I encourage you to think about it and to talk it over with close friends. I find that, when colleagues and patients of mine actually think this over, it tends to change their behavior—often in the direction of posting less and posting about fewer hot-button issues.

However, others I know have, after considering all the implications, doubled down on their posting; they feel solid in their beliefs and are not concerned about consequences. The point of this evaluation isn't to make one specific choice over another. It's to expand your understanding of how the medium affects others. This way, you can engage with social media in a way that suits and empowers you.

DEALING WITH SOCIAL MEDIA—AND ROAD RAGE

The other thing to consider when it comes to your negativity threshold is the amount of negativity *that you yourself are willing to experience*.

As chapter 15 explores, social media tends to be a lightning rod for negativity. In part, that's because of the situation discussed above: The chances are high that while scrolling through dozens—if not hundreds—of social media posts, some of those posts will upset us, even if they weren't meant to.

We also need to keep in mind that we have different negativity thresholds at different times. For example, we need to consider carefully how we use social media and other digital technologies during our more vulnerable times. When we're having a great day or are coming off a personal victory, small challenges or annoyances are easy to shrug off. But during challenging times, even a small perceived affront can easily overwhelm us (which chapter 38 discusses in more detail).

Even when we're *not* feeling vulnerable, we *will* experience negativity on social media. Given the sheer volume of material—not to mention the tendency for bad news and other information to propagate—it's a certainty. So, what do we do about this?

First, consider how much negativity you are willing to *take*. Those who answer zero have only one real choice: do without social media. If the answer is more than zero, you need to think carefully about your own threshold and learn ways to create a healthy dose of resilience in the face of negativity.

When we know what to expect, we can prepare ourselves and create a plan for how to handle it. One key strategy for deflecting negativity is knowing how to defuse situations on social media, even when people are hidden behind veils.

In person, you can read people's body language, and we are socialized to be polite when meeting others. But the "distance" provided by social media can lead people to be more aggressive and less controlled than they might be otherwise. It's kind of like road rage. A perfectly reasonable and ordinary person in a car, when provoked, might start shouting and gesturing in ways they never would if they were face-to-face with someone.

Why does the distance of being in a car bring this out? One reason is that driving is stressful. Whether we acknowledge it or not, every time we get behind the wheel of a car, we're operating thousands of pounds of lethal machinery, and at any given moment, we're potentially seconds from a nasty accident. While using social media doesn't risk our physical lives, there are more similarities than you might imagine. Most of the time on social media, we're sailing along and enjoying the ride. But if we accidentally post or send the wrong message, that could result in a potentially life-altering loss of a job, friendship, or reputation. This can substantially increase our underlying stress while online.

Another reason behind road rage is that people *dehumanize* one another. We yell, "That annoying blue car just cut me off!" Or we call the driver an idiot, a jerk, or worse. But it wasn't the *car* that upset us, it was a *person*. And if we met that person while walking down the street, or at a party, we would certainly be polite, if not genuinely friendly. Similarly, on social media, we see only people's veneers—their veils—so it is easy to dehumanize them. If someone's comment hurts us, we might trash them, thinking only of their confident, perfect-looking avatar and not of the vulnerable human behind it. This will be particularly the case if we don't know that individual personally.

Humanizing the people we interact with, and understanding that we all experience varied levels of stress on social media, can help us let negativity roll off our backs. It can increase the empathy and understanding we need to move beyond difficult situations.

This is a tricky balance. If we let *too much* roll off our backs, it can risk something else that is potentially problematic: desensitization.

In a remarkable study conducted at the University of California at Santa Barbara, researchers recruited male participants, who were told that they were jurors for a video trial. In fact, however, they were the *participants* in the experiment.

The pretend trial involved an alleged perpetrator and an alleged victim of sexual assault. Participants viewed sections of the trial, and between these sections they took breaks. During these breaks, some of the participants were shown violent "slasher" films in which women were victims. Other participants, however, were shown no content during the breaks.

Shockingly, compared with the participants who were shown no content during breaks, those who were shown the violent films became less sympathetic to the victim in the trial, less empathetic toward sexual assault victims in general, and less anxious and depressed at the horrible details of the rape described in the trial. Just by being exposed to brief portrayals of violence against women, they had become so desensitized that *their attitudes toward what they believed were real victims of violence changed significantly*.

This study is a cautionary tale. It is a reminder that, while we want to become more resilient, in some cases resilience can go too far. We don't want to become *too* good at handling negative material, since being exposed to and affected by so much negativity can end up desensitizing us to the pain of others. This is an important factor to consider when evaluating your own negativity threshold.

All of this can seem daunting, but for most of us, simply being more *aware* of the challenges will help. We can limit the power of inevitable negativity by humanizing others on social media. We can also learn to manage the stress we experience online by defining our negativity threshold and choosing to put down our phones or exit social media when we cross it.

Once we know where the line is, we can learn to trim more of the bad and bring forward more of the good.

17

Keep Your Feather Pillows Intact

THE 2008 MOVIE *DOUBT* CENTERS on the ambiguous relationship between Catholic priest Brendan Flynn and a student. Despite lack of clear evidence that the priest-pupil relationship is inappropriate, Sister Aloysius Beauvier spreads unfounded concerns about that relationship throughout the community.

At a critical moment in the story, Father Flynn tells his congregation the story of a nameless woman who has committed the sin of gossiping. Father Flynn describes how this woman is instructed by her priest to go to the top of her roof, cut open a feather pillow, and then return. She's confused, but she does as she's told.

When the woman returns to the priest, he tells her now to go back home and gather all the feathers that have been released. She of course responds that it is not possible to gather all the feathers, because they have scattered with the wind. The moral of Father Flynn's story is that the spread of gossip is like the spread of feathers. Once information has been released, the damage is done. There are many different versions of this story, including in Jewish folklore, but they all typically have the same message.

This story demonstrates the irreversible damage caused by gossip, which in our social media world can be increased exponentially. Since social media transcends geographic boundaries, the feathers of gossip don't just spread through a village and the surrounding countryside. In an instant, they can spread through an entire city and nation—and be transmitted to Australia, Algeria, Argentina, and beyond.

It's also *easy* to convey information through social media. We don't need to attend a dinner party to spread gossipy news or tell one person at a time. Now, gossip can spread at any hour of the day, from any location in the world, and to thousands of people at once.

As a result, people's lives can be dramatically affected in ways that were once impossible. I've had many patients in my medical practice over the years who felt that social media gossip made their mental health issues worse. While some people spread information on social media maliciously, more often the harm is inadvertent. A common theme is posted photos that accidentally show who was and was not invited to a gathering. No matter what the explanation turns out to be, the feathers have already been released. In extreme cases, affected individuals can experience depression and even suicidal thoughts.

We live in a society in which it can be hard to know what *is* gossip and what is not. Some consider gossip only to be malicious, harmful, untrue talk. However, others consider gossip more strictly; traditional Jewish views of gossip, for example, define it as *any* talking about others—regardless of truth or falsehood.

Complicating this is the fact that, in the world of "fake news," it can be difficult to identify what is real and what is not. To some extent, this has always been the case. The childhood game of telephone is an enduring lesson in how people can repeat information while causing slight amendments that result in an inaccurate picture.

But social media has changed the context. In today's world, this game of telephone can include billions of people. Advances in technology and countless news media outlets have made it extremely challenging to determine what is real and what is not. New artificial intelligence programs can make it look like politicians said something that they never actually said. And published research suggests that untrue rumors may spread just as easily—and for longer—than true rumors.

Despite these challenges, there are ways we can control the spread of gossip in the online world to make everyone's life a little easier.

IDEAS, EVENTS, AND PEOPLE

If technology can magnify gossip substantially, the first step to managing this is to recognize the issue and increase our sensitivity to it. In response, we can think more critically about spreading information—regardless of how "true" we believe it to be.

One concrete way to make this easier is to follow the advice often attributed to former First Lady Eleanor Roosevelt, though it was probably originally stated by historian Henry Thomas Buckle: "Great minds discuss ideas; average minds discuss events; small minds discuss people." If we follow this idea and allow it to change our discourse, we can elevate the value of social media in our lives while also reducing the risk of harm from gossip.

The second thing we can do is to recognize that—in those inevitable situations when we do spread information—*it can be less or more of a problem depending on the circumstances.* The scatter of feathers will be a lot worse in a windstorm than it is on a calm day. Therefore, we can be particularly sensitive when issues are more heated and have higher stakes.

Time also plays an important role. The gossiper in Father Flynn's story was so overwhelmed at the thought of capturing flying feathers that she didn't even reach out for one. Fortunately, there are dedicated tools for altering content on social media. Take the time to learn how to do this for each platform that you're on.

Snapchat, for example, is in "delete mode" as its default. Most messages automatically delete when they've expired or once they've been viewed. This sounds great. In a way our gossip automatically disintegrates, like the secret messages in *Mission Impossible*.

But remember that, once a piece of information is online, it often stays online. The original post may be gone, but the information can still circulate, digitally and in people's minds. Even posts that "automatically delete" can live forever: Posts are commonly saved manually by people taking screenshots. This is frequently how politicians and celebrities get into trouble. They might quickly realize that a recent post was offensive or easily misinterpreted, so they delete it. But inevitably someone has already taken a screenshot. Then the news article that comes out on *TMZ* or *Gawker* features the deleted post *and* an explanation of how the person quickly altered their feed. In some situations, this can look even worse than if the person had simply left up the post.

There are even computer programs that exist for the sole purpose of saving posts destined to be automatically deleted. Because people don't realize this, they often use Facebook to post benign content and a platform like Snapchat to post more risqué or otherwise questionable content. But that Snapchat post is quite possibly being saved somewhere, and it just might resurface during your Senate confirmation hearing.

Whether we like it or not, the spread of information on social media can be like cutting open a feather pillow in a windstorm. The second it's released, that information may be difficult or indeed impossible to retrieve. Therefore, it becomes particularly important to manage that spread—through sensitivity, caution, and wisdom.

This mindset can help us pivot from the natural state of social media to something that truly emphasizes the *positive*. If we're taking the bait and reposting rumors and gossip, we're not only part of the problem—we're also taking our energy away from the great positivity that social media use can be about: supporting and lifting up others, discussing meaningful ideas, and enriching our lives.

18

Balance Confidence and Humility

OPEN ANY INSPIRATIONAL QUOTATION BOOK, and you'll discover multiple philosophers, spiritual teachers, and leaders of industry describing why and how you should be self-confident. Two particular quotes have always stood out to me about confidence:

> **One is from the Dalai Lama:** "With realization of one's own potential and self-confidence in one's ability, one can build a better world."

> **And the next is from Allegra Versace:** "To me, sexy is a woman with confidence. I admire women who have very little fear."

As these quotations suggest, self-confidence is a sort of prerequisite to doing almost anything of great importance. If you don't actually believe you can do it, how are you going to convince anyone else that you can do it? How are you going to get the energy and positivity needed to put one foot in front of the other and complete the journey?

It's telling that both His Holiness the Dalai Lama and a celebrated socialite from the world of fashion design share their belief in the importance of confidence—even if they do have somewhat different reasons for their interest in it.

However, as you turn the page in your inspirational book of quotations, you'll probably also see a quote about humility. You may see this classic line from Thomas Moore: "Humility, that low, sweet root, from which all heavenly virtues shoot."

Or this one from Simone Weil: "Real genius is nothing else but the supernatural virtue of humility in the domain of thought."

These quotations also make sense. Moore was born into an Irish Catholic family of limited means. But through his intellect (and perfection of an English accent), he was able to get into Trinity College—which had only recently started admitting Catholic students—and forge a successful career as a poet and singer. Remembering his humble roots helped him to navigate, and then thrive in, his new upper-crust environment.

Weil's situation was also complex. Born in 1909, she was a genius and philosopher who spoke multiple languages, including the ancient Greek she learned as a child and the Sanskrit she learned in order to read Hindu religious writings in their original language. As a passionate altruist and dedicated political activist, she is said to have starved herself to death in 1943 at age thirty-four in solidarity with Holocaust victims. Although she had tuberculosis and was living in the United States, she reportedly did not want to eat any more than Jews were allowed to eat at the time in Nazi-occupied France. While you might say that her humility cost her her life, it also formed the basis of the success in her life—her altruism, her passion, her search for knowledge, and her legacy.

So, who's right? The Dalai Lama and Allegra Versace or Thomas Moore and Simone Weil? Should we aim to be confident or humble?

Well, both, of course.

Even though words like *confident* and *humble* are often considered opposites, this is one of those cases in which they're not so much antonyms as related attributes on a continuum. I propose a slightly different picture of the relationship between confidence and humility—one in which our perspective makes the difference (see figure 3).

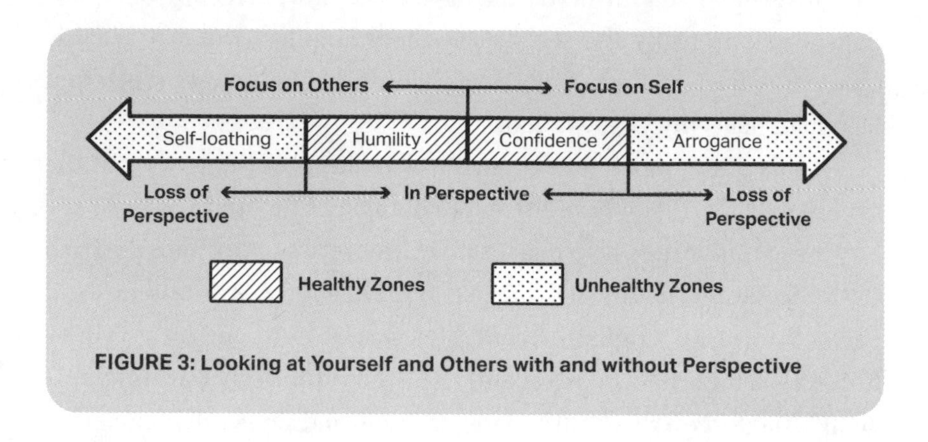

FIGURE 3: Looking at Yourself and Others with and without Perspective

Confidence and humility are strongly related, but one arises when we focus on self, and the other arises when we focus on others. However, each attribute can become unhealthy when it becomes too extreme, or when we *lose the perspective* of the other attribute. In the case of confidence, we focus on our worth as a remarkable human being—the fact that we have tremendous potential and deserve compassion and opportunity. But when we lose the perspective of the importance of *others,* confidence can veer into arrogance—the belief that our needs and value are somehow more important than everyone else's.

In the case of humility, we recognize that we are just one person within a much bigger world and that all people have value and importance. But the risk here is that, if we lose the perspective of *our own* value and worth, healthy humility can devolve into self-loathing.

What does all this philosophizing have to do with social media?

A lot, I think. Next time you log on, notice that much of what goes on in the world of social media is a continuous dance between confidence, humility, arrogance, and self-loathing.

We need at least some level of confidence to post anything at all. Every time we post we're likely to ruffle some feathers, even if we don't mean to. So, to put ourselves out there in the first place, our self-confidence needs to pass a certain threshold.

I encourage you to reach into that confidence when you use social media. After all, it's out of that confidence that His Holiness the Dalai Lama and Allegra Versace encourage us to create positive change and empower others to live with fearless self-expression—things that certainly can happen online. This includes encouraging others during their challenging times, magnifying their accomplishments and good feelings, and engaging in valuable self-expression.

However, we've all seen it go too far. We've seen people on social media who can't stop talking about themselves. This happens in real life as well; some people will dominate conversation and demand everyone's focus. But there's something about social media—and other digital technologies like Zoom—that accentuates this dynamic. At least if we're talking in person, we might hear our own voice and realize we're dominating the room. The next time, we might pause before presenting a dissertation on why we chose one toothbrush over another. But in the world of social media, those auditory cues are often missing. It's also not easy to read the body language of others, even on a visual platform like Zoom. So, online, it becomes that much more important to temper our confidence, keep it in perspective, and not let it fall into arrogance.

Social media can also breed arrogance because of its connection with celebrity culture. Sports figures and actors have always been—and remain—prone to an exaggerated sense of self-importance. Now the role of "social media influencer" is suddenly on par with these other professions. A 2018 report suggested that an individual with over a million followers can command as much as $100,000 to $250,000 *per post*. This has led to an arms race in which the number of followers, friends, or fans someone has equates to their value as a human being—and this includes all levels of social media users. It's a trap we all need to be mindful to avoid as we try to maintain humility.

However, having humility on social media doesn't mean that we should disappear into self-worthlessness, either. As chapter 26 explores, being a passive wallflower can foster things like depression, anxiety, and loneliness. How do we find and keep the right balance?

One practical and concrete way is to look back over and consciously think about the frequency and content of your posts. That pattern itself can reveal things you didn't realize at the time. Did you get too caught up in yourself after a particular success? Were you posting about the same event—with unwavering frequency—months after the fact? On the other side, did you retreat into silence for too long after one disappointment?

When things in life seem too chaotic on a broad level, we can find solace in dealing with it in a small, metaphoric way—cleaning out our desk, bedroom, or car, for example. Toward the end of the movie *Crazy Heart*, Jeff Bridges's character, an alcoholic country star, cleans his home as a way of getting himself out of a vicious spiral of addiction and loss.

In the same way, analyzing and adjusting your social media life can help you attack larger life issues around confidence and humility. A slow foray back into posting can help build confidence, which can bleed into your offline life. After a break from social media, if you start to comment on others' Instagram photos, to connect with them one-on-one to see how they're doing, and ultimately to post some of your own material, it can help pave the way for making similar moves in the offline world. "Practicing" the skills involved in reaching out for help on social media during hard times—and going through some of the associated emotions—can transfer onto other situations. In this way, social media can be a tool to help build confidence when you need to do that.

On the other hand, this same kind of self-reflection can lead to noticing that you're focusing too much on trivialities in your life without paying attention to what others are going through. This insight can help you develop more compassionate humility when dealing with others.

This process can be more challenging than it sounds. It can be hard to find the time in the first place to consciously look over your feed.

Then it can be hard to take objective stock of what you see. Finally, there's the challenge of creating a new plan and sticking to it.

For this reason, I suggest doing these things *with someone else*. Another person can help you carve out focused time. Looking over each other's feeds will help you face difficult realities together. And it's easier to stick to a plan for change when you have someone providing support.

Of course, choosing the best person for this can also be difficult. This exercise can get quite personal, so it's important to share with someone you feel comfortable being vulnerable with. Almost like the quarantine "pods" that developed during the Covid-19 pandemic, form this relationship with someone you trust emotionally.

Finally, this exercise of balancing confidence and humility comes back to the second level of the social media pyramid—*being positive*. In figure 3 (page 112), there is another important difference between the two middle attributes—being confident/humble—and the extreme attributes, being arrogant/self-loathing. It's *positivity*. Both confidence and humility are about consciously *looking for the good*.

When it comes to confidence, it's about looking for the good in yourself and celebrating that. There is nothing wrong with touting our accomplishments, posting pictures of ourselves that capture our individuality, and sharing our valuable thoughts. When it comes to humility, it's about looking for and amplifying those good things in *others*.

At any time, in any situation, whether on social media or offline, it's a good idea to be positive—and to let that positivity keep you in the wise continuum between confidence and humility.

19

Be Positive with a Vengeance

DESPITE HOW CRITICAL POSITIVITY IS, we also know how challenging it can be to maintain on social media. After all, look at what we're up against—negativity bias, toxic gossip, dehumanizing language, extreme reactions and behaviors, a set of rules that seem to facilitate the spread of negativity, and more.

That's why the critical message of this chapter is that we can't just be "sorta positive." We need to actively search for it and *pursue positivity with a vengeance.*

In the face of difficult experiences on social media, even when we're being criticized and feeling irritated, offended, and possibly horrified, we should try to respond with acceptance and composure.

How? The answer comes from Viktor Frankl, one of the most important psychologists of the past hundred years. Frankl was a Holocaust survivor who watched while his friends and family were killed and tortured in concentration camps. Yet, remarkably, he focused on the small acts of light and kindness he saw. He noticed that one person who was starving to death would often give his last crust of bread to another person who was also starving to death. He was amazed by the fact that these enslaved people exercised their ultimate freedom in the way they *responded* to their situation—they refused to devolve into animosity and selfishness.

Frankl's ideas can be summarized in the quotation for which he's best known: "The last of human freedoms is to choose one's response to any situation."

If those in the worst possible conditions can choose to find good in such darkness, we can learn to seek out the good in our everyday lives. We can do it with our work, our relationships, and on social media. This doesn't mean that we compromise our values and that we never respond in a negative way to anything. It just means that we *focus* on positivity.

CLINGING TO OLD AND ABSENT FRIENDS

Another quotation when considering how to actively pursue the positive on social media comes from a fictional character—Katya Kontent in Amor Towles's *Rules of Civility*.

Katya, the daughter of a Russian immigrant, uncovers hypocrisy and darkness beneath the rosy exteriors of the upper ranks of 1930s New York society, yet she maintains humanity and positivity. At one point, despite being surrounded by many people who all have ulterior financial and political motives, she recalls a meaningful friendship and thinks to herself: "When some incident sheds a favorable light on an old and absent friend, that's about as good a gift as chance intends to offer."

In addition to being a warm and positive sentiment, this statement gives us a practical clue for using social media in a positive way in today's world.

One of the best things about social media is the ability to stay connected to "old and absent friends." But how *specifically* should we use social media to connect with these people so that we truly optimize those relationships—and avoid falling into misunderstandings and other traps we've been discussing?

Katya lives in a time without social media. She must wait for a specific happenstance to shed a favorable light on a former friend—maybe hearing a name in passing or coming upon a physical reminder while going through old things. Without these reminders, many wonderful old friends might remain unremembered—which would be Katya's loss.

Yet with social media, we have these reminders available to us *every day, every moment*. We can experience the warm feelings Katya refers to frequently. We can use social media to choose the company we want to keep.

All we need to do is to log on to our platform of choice, see what's going on, comment on those people's posts, and reconnect, right?

No. We must dig beneath the surface to have a richer experience. We must pursue positivity *with a vengeance* instead of just waiting for it to happen. When we log on to the main feed, we are hearing the loudest voices. These may or may not be the ones that we want to hear. We're also getting fed the voices *selected by the platform*—voices selected to be the most fulfilling for the platform and not for us.

Imagine that two messages are posted at the exact same time. The first is from an old friend we miss and haven't heard from in a while—it's not flashy, so it doesn't get a lot of hits. It also doesn't have any words in it that suggest a product that could be sold, like *computer* or *car*. The second message is from someone who posts frequently and carefully curates messages to get maximum hits. Platforms are often optimized to emphasize messages like these because they will end up selling more or keeping more people glued to their feeds. When there are a lot of hits, algorithms tend to look for key words in the message. If the message is about a great camping trip, an ad for a camping store will probably appear next to it. This isn't just the case on Facebook—similar algorithms and processes are likely at work on Twitter, Instagram, and just about every other social media platform.

You know what happens. The platform will usually put their preferred messages at the top of your feed, and other messages, the ones we probably want and prefer to see, will appear farther down. What can we do about this?

We need to be more *intentional*. With a vengeance, even. We need to have a plan that involves skipping the highly visible screens, which platforms default to for their benefit. Instead, we need to drill down to what we want. Here is one example of a practical way to achieve a more tailored, personal, and intentional experience.

Bookmark the *personal pages* of five to fifteen people that you specifically want to know about, and make shortcuts on your desktop or device directly to the pages of those people. Then, when you have a break and want to check social media, go directly to those spaces that you *truly want to inhabit*. This will be more beneficial to you compared with just entering the platform's grand marketplace, which is likely to carry you away to less-relevant, less-personal, and more-commercial information. This way, you go directly to the old and absent friends you want to interact with.

You can develop your own system over time. If you start with ten individuals, a few weeks later you might add on a few more. On the other hand, after a few weeks, you may realize that you mostly focus on seven people, and so you remove the other three and hone your list.

This technique gives you one practical way to *curate* your immediate environment. It can make it more likely to find the old and absent friends that you're truly looking for. We can't expect optimal interactions to happen without intention. We need to identify the positivity we want—including the specific memories and exchanges we crave—and then pursue them with a vengeance.

20

Dance Like the Whole Universe Is Watching

BUMPER STICKERS AND INSPIRATIONAL BLOGS are emblazoned with the phrase "Dance like nobody's watching." The quotation has sometimes been attributed to writer and humorist Mark Twain. Other people claim that it was spoken by American baseball star Leroy "Satchel" Paige. However, there's little actual evidence that either of these men ever said it.

The first time these words were published was not until 1989, when Kathy Mattea, a country star, recorded "Come from the Heart." Then William W. Purkey, a professor of counseling education, adapted the chorus of that song into a poem:

> You've gotta dance like there's nobody watching;
> Love like you'll never get hurt.
> Sing like there's nobody listening;
> And live like it's heaven on Earth.

Regardless of where the line comes from, it has survived because it captures the imagination. There's something compelling and freeing about it. Don't we all feel like letting go of societal customs and traditions at times? This sense of liberty and spontaneity has sparked creativity throughout history. It releases people—even if just temporarily—from the doldrums of our monotonous routines.

The problem is that—in the age of social media and other digital technologies—the consequences of spontaneity have changed. In the past, if you let loose at a party and started dancing on the table and belting out an off-key version of "Take Me Home, Country Roads," this would become a funny story to tell among friends, but it wouldn't be recorded. Now, anything can be captured, turned into a viral meme for the ages, and shared with the world. Further, whatever happened can appear out of context, which can make it easy to misinterpret.

Here is one mild, ostensibly harmless example. In early 2019, a man was recorded on a stranger's phone as he dragged his daughter through an airport luggage-return area by the hood of her jacket. She was lying on the ground as he pulled her along, and both of them were quite calm, as if they'd done this a thousand times. Before the internet, passersby would have noticed, raised a few eyebrows, and probably wondered if this was okay or not.

Today, however, this episode was recorded and posted to a YouTube channel. It promptly went viral. Over two million people viewed this event, and such a swirl of commotion arose that it became a news story. Was this an abduction, abuse, or just a clever father successfully managing his bored kid?

The truth is, we can't know. The video seems to capture an innocent moment, but taken out of context, viewers can't confirm what is actually happening. Even if the man eventually realized he'd become famous and posted the real story in the comments section, how many viewers would ever see that explanation? Meanwhile, the meme lives on.

Indeed, some people use the internet and the ability to record anything to help bring to light terrifying episodes of violence and brutality. The Reddit community "PublicFreakout" has captured numerous episodes of inappropriate abuse, and it currently has millions of subscribers.

Today, whether in public or private, we're always on the verge of being recorded for posterity. But people don't usually get into trouble over viral videos taken by someone else—most often trouble arises because of their own words and posts. On its website, one major New Jersey personal-injury law firm offers advice for what to do "if a comment on social media gets taken out of context resulting in an arrest." That's right—people can be arrested for things that they express on social media. While that's relatively rare, it illustrates the real-world consequences of social media. Our comments not only can be misinterpreted by our friends, family, and acquaintances but also can lead to unwanted public infamy and even legal jeopardy.

Thus, while we might want to "dance like no one's watching," when it comes to social media, I suggest dancing like the universe is watching, because every keystroke you make can follow you in perpetuity.

On top of this, social media posts are often treated like public writing and reviewed by employers. For example, just about every university vets future employees and students by reviewing their visible social media activities. Applicants are commonly rejected from colleges because of questionable things said on social media. How many of these rejections are deserved and legitimate, and how many take social media posts out of context? Job applicants are also now commonly rejected because of their social media presence, and one of the top reasons for letting employees go has become because of social media postings.

And it's not just professional contacts who vet us. People we meet at work, new friends, and potential romantic partners can and do review our social media presences to find out who we are.

So, how do we achieve that complex balance between freely expressing ourselves and proactively censoring ourselves, since every comment on someone else's comment might end up costing us our dream job?

One thing to keep in mind is that it's not always as bad as it seems. Because of the sheer mass of information online, many embarrassing moments really will get swallowed up and forgotten over time. I'm reminded of the 1979 Carl Reiner movie *The Jerk*. At one point, the main character, Navin R. Johnson, becomes frantically joyful to find his name in the phone book among millions of other names. With glee and anticipation, he declares, "I'm in print! Things are going to start happening to me!"

The film immediately cuts to a sociopath sniper holding a rifle and deciding whom to kill. The sniper closes his eyes, randomly points to a name in the phone book, and happens to settle on Navin R. Johnson.

You get the message. Being "in print" *could* be dangerous, but the likelihood of serious trouble is probably minimal. The analogy isn't perfect, because the internet is quite different from a phone book. But the movie's point is well taken: While we need to be thoughtful and intentional about social media, just because we post doesn't mean that everyone on Earth will see it, remember it, or care.

POSITIVITY IS THE ANTIDOTE

The most powerful weapon we have if we want to dance as creatively and whimsically as possible—even with the understanding that everyone's watching—is being *positive*.

If we are positive with our actions, our posts, and our shares, we will avoid the vast majority of potential pitfalls. It's generally the negative, abrasive comments that raise eyebrows and get propagated. It's generally the negative things that we do in public that cause problems. Just ask athletes and celebrities.

Certainly, posting on social media comes with risks. But here are four suggestions that constitute a *positivity antidote* to help avoid social media pitfalls.

If you have to ask the question, the answer is often no. If you are not sure of the appropriateness of what you're about to post, it probably isn't the best idea. Because of the sheer number of people following you, their extended connections, and those people's connections, if you're second-guessing something you're about to say, someone else will probably be troubled also. In these situations, you don't have to completely censor yourself. One solution is to share the information in a more limited way—like in a group email, a private message, or a phone call.

Take your time. Many people now make it a practice to have their email automatically delay sending their messages. You can set most email programs to automatically delay each transmission by an amount of time you select. After you press Send, a button appears that says something like "Undo Send"—and you can then press that if needed. This is not only useful if you've accidentally said something that you want to reconsider. It's also helpful if you realize you forgot to add an important detail or an attachment. I understand that the world of social media doesn't often work this way: It's designed to be immediate. If you delay sending, even by minutes or seconds, you might feel like you're "behind." Nevertheless, I think it's worth it on balance to sit on what you've written for a bit. Maybe you'll send it as is a few minutes later. But maybe a spark of positivity will inspire you to make some slight tweaks that ultimately improve its reception.

Check with someone else. Social media, interestingly, is often a highly *solitary* activity. Yes, you're "interacting" with 742 friends. But ultimately, if you think about it, it's just you, by yourself, hunched over, thumbs punching away. Ironically, any opportunity to make social media a bit more "social" is welcome. You can get someone's opinion from afar, but having them give you advice at a coffee shop or in another in person setting is even better.

Don't post anything you wouldn't want your grandmother to see. This is one I often tell young people who are just getting started with social media. Admittedly, we all have different grandmothers who have various levels of comfort with different activities and ideas. But nonetheless, this rule often seems to help people remain positive and avoid challenging situations.

In summary, don't think that nobody's watching on social media—they are. While this can come with challenges, armed with an antidote of positivity, we're ready to reach the peak of the social media pyramid. It's time to get creative.

Part 4

Be Creative

THE WORD *CREATIVE* OFTEN MAKES people think of something like "being good at arts and crafts." But for our purposes here it's broader and bigger. Creativity in this context is more about being *original*, *proactive*, and *open to change.*

Interestingly, the use of the word *creativity* is a modern phenomenon. The word was barely used in the early 1900s, but its use in English-language books skyrocketed more than twenty-fold between 1940 and 2000. This parallels the importance of this value in modern-day society.

In my previous position as a medical school administrator, I interviewed candidates to be future doctors, and a key attribute we looked for in applicants was creativity. In the past, medical schools looked for people who could remember a lot of facts and had scientific experience. Now we recognize that people who are more original, well-rounded, and flexible are the most likely to succeed as medical students, physicians, and physician-scientists.

Creativity belongs at the pinnacle of the social media pyramid because it helps you optimize your use of social and digital media. Here are three reasons why.

First, creativity—being original, proactive, and flexible—is a way of taking something *potentially problematic and turning it into something valuable.* When someone gets booted from a job unfairly, one path is relatively standard: moping, anger, and settling for the next position that comes along. But the more creative path involves thinking outside the box, being proactive, and finding something even more fulfilling.

We already know that social media can present difficulties. We know that heavy social media use is linked to emotional health problems. We know social media platforms are designed to be sticky; they manipulate our attention to keep us clicking. To one degree or another, we've experienced these things for ourselves.

That's where creativity comes in. We need to be inventive, original, and proactive to take those challenges and turn them into positives.

Social media marketers have been known to use the strategy of *removing friction*. In physics, friction is the force that slows things down. It's the reason that a sled sliding across level snow eventually stops instead of sliding forever. In the case of social media platforms, an explicit goal is to reduce to a bare minimum any "friction"—the force that leads us to stop clicking and get back to our lives.

Creativity is a tool we can use to *reintroduce friction*. By being original, proactive, and flexible, we can convert something that drains our attention into something that enhances our lives.

Another aspect of creativity involves *tailoring*. This means analyzing the specifics of a situation—and then thoughtfully selecting the right tools and techniques to get the job done. This kind of creative tailoring is critical to making sure that your digital life works for you instead of against you.

Tailoring in other contexts is known to improve effectiveness. Advertisers don't create one ad that's meant to apply to everybody. Instead, they carefully segment their market into multiple niches and create targeted advertising based on demographics and personality characteristics. For example, one major tobacco company categorized their potential markets of young adults as either "progressives," "Jurassics (conservatives)," or "spoiled brats." Then they created different campaigns aimed at pushing different brands—in specific ways—to each of these groups. They were highly successful.

Public health professionals use these same techniques to *improve* health. A one-size-fits-all approach is not optimal for behavior change. Tailoring helps reach specific people in specific circumstances to show them how to do things like avoid substance use, take medicines on time, get required health screenings, and follow appropriate diet and exercise regimens.

Similarly, we can creatively tailor our social media use to our own *personalities*, which will help us have much richer experiences. Instead of following the crowd—and what marketers want us to do—we can take control of both our health and our attention. This is one reason why it's hard to give blanket recommendations for complex behaviors like social media use. Instead, throughout part 4, I suggest that you take the general advice and *tailor it to your style of social media use and your personality.*

Finally, creativity belongs at the pinnacle of the social media pyramid because it is one of the most *central aspects of being human*. That might sound dramatic—but it's true. More than any other animal species, humans are defined by the ways in which we think critically, reflect, change our behavior, and optimize our lives. And we can use those abilities to consciously and proactively craft a social media plan that serves us.

The pinnacle of the pyramid is the hardest part, but it's the part that will most dramatically help you to leverage the way you use social media to forge your own future.

As a first step in that direction, please complete this brief personality quiz on the next page, which you can use to help you hone *your* personal plan.

Personality Quiz

I based the questions in this quiz on established personality research. Understanding more about your personality traits will help you tailor your social media use to your life.

First, in table 1, put a number between 0 and 5 in each of the blank boxes based on how strongly you agree or disagree with each statement. Put 0 for strongly disagree, 5 for strongly agree, or any other number in between.

Be honest with yourself. Don't respond with how you *would like* to be—respond with how you honestly feel you *are* at this point in time.

Table 1: Personality Quiz Statements

Statement	Do you agree with this statement (0-5)?
1. I see myself as extroverted and enthusiastic.	
2. I see myself as critical and quarrelsome.	
3. I see myself as dependable and self-disciplined.	
4. I see myself as anxious and easily upset.	
5. I see myself as open to new experiences and complex.	
6. I see myself as reserved and quiet.	
7. I see myself as sympathetic and warm.	
8. I see myself as disorganized and careless.	
9. I see myself as calm and emotionally stable.	
10. I see myself as conventional and uncreative.	

Next, in table 2 below, fill in your score (in the right-most column) for each personality characteristic using the numbers you wrote in table 1. Use the formula in the middle column to calculate each score. For example, to figure out your score for "conscientiousness," take 5, add the number you put down for statement 3, and then subtract the number you put down for statement 8. The total score for each personality characteristic should be between 0 and 10.

Table 2: Personality Quiz Scores

Personality characteristic	Formula for your score	Your score
Conscientiousness	5 + statement 3 – statement 8	
Agreeableness	5 + statement 7 – statement 2	
Neuroticism	5 + statement 4 – statement 9	
Openness	5 + statement 5 – statement 10	
Extroversion	5 + statement 1 – statement 6	

These particular personality traits—conscientiousness, agreeableness, neuroticism, openness, and extroversion—are often regarded as the "big five," and the next chapters examine each in more detail, along with what various scores mean. In the meantime, you're ready to start tailoring your social media use to your personality.

21
Consciously Conscientious

IS IT GOOD TO BE conscientious?

I realize that sounds like a trick question. Most people would likely think: *Of course it's good to be conscientious!* People who are conscientious tend to be more organized, dutiful, and reliable. Taken to the extreme, though, conscientiousness can also be a drawback. People who are highly conscientious also tend to be perfectionists and inflexible.

An appropriately conscientious coworker might make your life significantly easier by having a report on your desk two days early and with no errors. A coworker without conscientiousness might forget and never deliver the report at all. But an *overly* conscientious coworker might get irritated at *you* for not finishing *your* part of the project early, even if you are still on track to complete it on time.

Most of us know people at different places along the conscientiousness spectrum. How do you think social media might affect their emotional health differently?

One of my students, Erin Whaite, and I explored this question by testing the relationships between personality characteristics like conscientiousness, social media use, and loneliness.

We found interesting results for conscientiousness in particular (see figure 4). For people who tested *high* in conscientiousness, there was no relationship between social media use and loneliness.

However, for people who had *low* conscientiousness, there was a dramatic relationship: In this group, compared with those who used the least social media, people who used the most were more than three times as likely to feel lonely.

On the personality quiz (see page 132), how did you score for conscientiousness? If you scored high, our study suggests that you are at a lower risk of feeling lonely, even if you use a lot of social media. If you tested *low*, then the more you use social media, the higher your risk of feeling lonely.

Figure 4 does *not* mean that conscientious people use less social media. Instead, it means that conscientious people are somehow able to *handle* the experiences they have on social media better. This is heartening because it means that we *all* might be able to learn something from conscientious people to improve our social media lives.

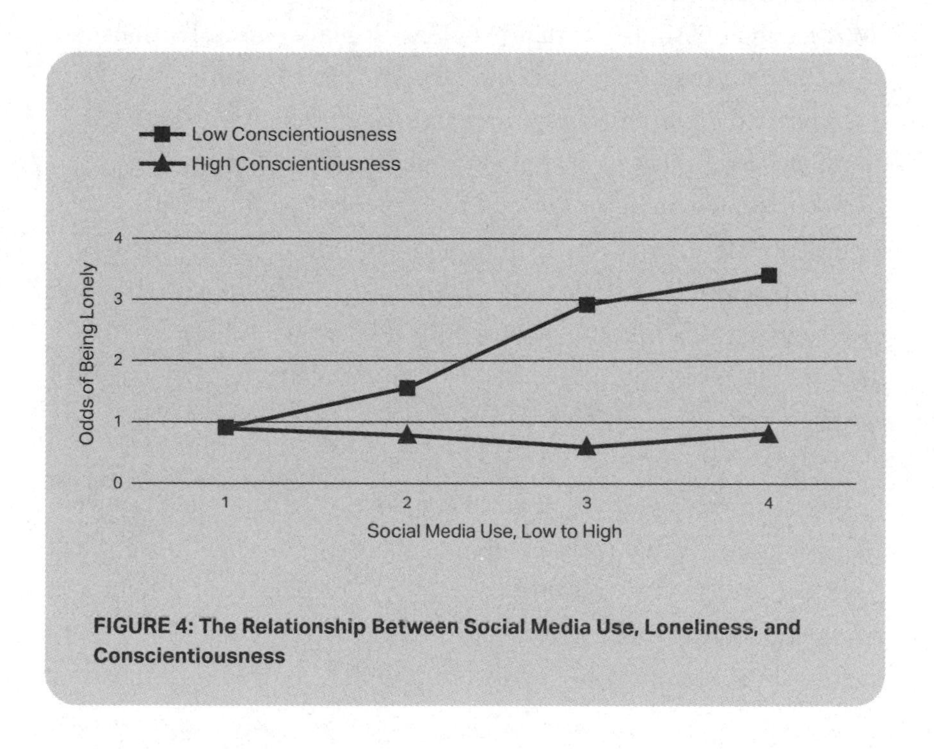

FIGURE 4: The Relationship Between Social Media Use, Loneliness, and Conscientiousness

So, what is their secret? How is it that highly conscientious people seem to be more "immune" to heavier social media use?

One commonsense possibility is that conscientious people are likely to make thoughtful choices when posting and commenting. A conscientious person's tendency to be mindful and respectful with how they treat others—in any context—creates less drama. It may seem obvious or simplistic, but being intentional likely reduces emotionally difficult situations and experiences.

Conscientious people may also be *less impulsive* when posting or commenting, lowering their risk of making a gaffe. As chapter 20 discusses, social media "fails" are common and potentially dramatic. Fails can lead to losing jobs, friends, romantic relationships, and dignity. Obviously, these kinds of losses can leave us lonely and depressed. Being more aware of and sensitive to others can protect us from these kinds of situations.

A second possibility is that people with higher conscientiousness might *balance* their online and offline worlds in a healthier way. Even if they spend a fair amount of time online, conscientious individuals might make it a point to keep their in-person social lives active.

For instance, imagine two different people who both use social media three hours per day. The person with more conscientiousness might use the rest of their time to schedule in-person activities and to meet friends, while the person with less conscientiousness might neglect this. That could certainly lead to the pattern my team and I saw, in which people with low conscientiousness were at higher risk of loneliness when they used a lot of social media.

Don't worry if you didn't test very high on this personality trait. The good news is that you can learn to act more conscientiously. Each time you draft a post, remind yourself to consider carefully how it might make others feel, and revise it as necessary. Each time a comment or post angers you—instead of impulsively replying—channel your inner conscientiousness, pause, and contemplate whether a full-out debate on social media is truly constructive and worth your time and energy.

On the other hand, if you tested high for conscientiousness, don't regard that as permission to gorge on social media; the other lessons in this book still apply. As general advice, it's healthier for everyone to take a personal inventory of their social media use and make sure they spend at least as much time with friends, family, and loved ones in face-to-face connection. This is simply a conscious and conscientious way to bring balance to your life.

22

The Double-Edged Sword of Agreeableness

THE NEXT "BIG FIVE" PERSONALITY trait to consider is "agreeableness." How did you score? Do you tend to be more trusting, forgiving, and straightforward, or more tough and suspicious of others?

There's no "right" way to be. For example, people on both sides of the agreeableness spectrum have the potential to be great leaders—they're just *different kinds* of leaders. People who are agreeable tend to be more "transformational" leaders, while people who score lower on agreeableness tend to be more "transactional" leaders. Transformational leaders work with teams to create shared visions and strategies, while transactional leaders focus more on organization, rules, and performance-based measures. Both kinds of leadership are important, and either one can lead to success. Often a combination is valuable.

When studying this personality trait with my team, we discovered that those who are more agreeable were less likely to feel lonely in general. After all, being agreeable means that you are more likely to be positive and forgiving. This may make you less prone to engender conflict in your relationships.

When we studied how people with this personality trait reacted to social media, we expected those with high agreeableness to also be less *susceptible* to the negative effects of using a lot of social media—like we saw with conscientiousness.

Interestingly, that wasn't the case. We found that more social media use translated into more of a feeling of loneliness—*regardless of someone's level of agreeableness.*

One reason for this might be that agreeableness can be a double-edged sword. People with high agreeableness are often better at letting more roll off their backs. On the other hand, these people also may be more likely to get overlooked or marginalized on social media. Their tendency to agree might end up making them feel less central to the discussion, which can make them feel lonely.

What about someone with low agreeableness? Because they are tough-minded and determined, they might be less sensitive to difficult things that happen to them on social media. Let's say this type of person streams a live Instagram video of themselves performing, on an old acoustic guitar, the arpeggiated opening of Led Zeppelin's "Stairway to Heaven." Not getting likes or comments on this might crush some people, but a person with low agreeableness might shrug it off.

On the other hand, people with low agreeableness, or who aren't afraid to have a different opinion, can come off as brash and argumentative. This can lead to being unfriended, ostracized, and ultimately more isolated.

How might your agreeableness score affect your social media use? Either way, it might seem like you're fighting an uphill battle: too nice and you're a pushover, too disagreeable and you're pushy. The solution comes back to recognizing the benefits and drawbacks of each tendency. Then you can bolster the benefits and be cautious about the drawbacks.

If you tend to be more agreeable, take advantage of that by supporting others on social media; go with your natural desire to be helpful. But also remember that your kindness and optimism might not always be reflected back to you. It can be surprising and disturbing for agreeable people—especially when they are just starting out on social media—to find that people don't treat them the way they tend to treat others.

If you're agreeable, it can also be difficult to have less of the spotlight. But remember that, even if you're not always center stage, you can still find ways to express your opinions and gain confidence and fulfillment from that. For example, you can focus your communications on individuals or smaller groups of people who will be more likely to match your enthusiasm and sensibilities.

REMEMBER THE HUMAN

If you are lower on the agreeableness scale, then in a way you've got an advantage—your toughness and determination may carry you nicely through some social media challenges, like the awkward silence you felt after the debut of your "Stairway to Heaven" masterpiece. You may also be more in the limelight, because you're not afraid to express your opinions. But know that the tendency to stand up for something strongly can also lead to hard feelings.

This can especially be the case in the milieu of social media. In person, you can disagree with someone, but your smile and tone can show that you're still being respectful. On social media, tone is harder to convey. People don't "hear" what you mean, they hear what they hear. This is true in real life, but even more so on social media because of the lack of certain verbal cues.

One solution for this is to liberally use emoji (see chapter 29 for more on this). No, a pixelated yellow blob with fake eyes and a fake smile isn't the same thing as a true human expression. But if you're concerned about softening something that could come across as harsh or be misinterpreted, that blob could do the trick.

It is always important to "remember the human" online, which is one of the major guidelines for the Reddit platform. On platforms like Facebook, user transparency is required. You are supposed to have one account that's definitely you, so people use their real names the vast majority of the time. That's not the way it is on Reddit, where most people hide behind pseudonyms and often exhibit little concern for antagonizing other people. The result is a community that has advantages—like interesting, provocative, and well-organized content—along with challenges, such as less agreeableness. This is why Reddit finds it necessary to post the rule "remember the human," which is not bad advice for anyone on social media, regardless of personality.

23

Embrace Your Neurotic Side Before *and* After Posting

THE WORD *NEUROTIC* OFTEN CONJURES up a lot of negativity. You might picture Sigmund Freud interviewing a deeply disturbed patient, or Woody Allen anxiously babbling on a psychiatrist's leather couch. Or you might think of Riggan Thompson, the main character in the movie *Birdman*, consumed by debilitating distress. In the thesaurus, synonyms for *neurotic* include "mentally ill," "mentally disturbed," and "unstable."

Believe it or not, however, being neurotic is not as bad as it sounds. First, those synonyms in the thesaurus are simply wrong. Neuroticism as a personality trait is *not* the same thing as having a mental health condition. Psychologists specifically distinguish neuroticism from psychosis—which *does* indicate a break with reality and *is* a mental health condition. Neuroticism is strictly a personality trait that is better defined as "having a higher sensitivity to negative outcomes."

Sure, having that kind of sensitivity can feel uncomfortable. People with more neurotic personalities tend to worry a lot about things beyond their control—to the point that it causes undue distress.

But like a lot of other potentially negative traits, it also comes with possible advantages. In fact, being neurotic may actually come along with a survival advantage. In the ancient world, being more neurotic helped you prepare better for dangerous situations, like a woolly mammoth attack. In the modern world, people who are more neurotic are probably more likely to make sure they finish that homework assignment or that work report.

In our studies, we found that—without considering social media use—people with high neuroticism were more likely to feel socially isolated. And the relationship was strong. Compared with those in the lower half of neuroticism, those in the upper half were more than three times as likely to feel lonely.

What happens when you add social media into the equation? Interestingly, we didn't find that people on the more neurotic end of the spectrum were more *sensitive* to using a lot of social media. Instead, we found that *regardless* of the amount of neuroticism someone had, their loneliness went up when their social media exposure went up.

Why might this be? It's probably because having neurotic tendencies introduces both advantages and disadvantages when navigating social media. If, by definition, you have "a higher sensitivity to negative outcomes," then you may have learned to protect yourself by being more cautious. You might then be less likely to impulsively post that picture that someone might find offensive. You may also be less likely to fire off that angry retort when someone posts something infuriating.

Then again, if you're highly sensitive and something potentially embarrassing is posted about you, you may take it harder than other people would. And your concerns about that potentially embarrassing post or picture might *linger* for longer. Months after everyone else in your social media circle has forgotten about the "crazy hair" photo from your high school reunion, you may still be kicking yourself for not bringing a comb.

These examples suggest a clear path to improved use of social media for people with neurotic tendencies. Notice that the advantages are related to *before posting*, but the disadvantages are related to *after posting*. So, what people with neurotic tendencies can do is to embrace their natural tendencies *both before and after they post*.

Before posting, listen to that voice that is "sensitive to negative outcomes" and let it help you make better choices. Then, after posting, resist tendencies to fret about what can't be fixed. After that post is out in the world, it is important to shrug off that natural tendency to worry—especially when there's nothing you can do about it.

Like every personality trait, having neurotic tendencies can be both challenging and valuable. By virtue of being human, we all have a bit of every tendency. So, let's use them as best we can.

24

The Many Meanings of "Openness"

AS IT TURNS OUT, OPENNESS as a personality trait is hard for people to describe.

In our study on the relationship between social media and personality, it was also the most difficult trait to measure. To quantify openness—or what psychologists more completely sometimes call "openness to experience"—we asked people to agree or disagree with statements like whether they thought they were "original," "sophisticated in art, music, or literature," "inventive," "a deep thinker," and things like that.

These qualities are more abstract and ambiguous compared with some of the other "big five" characteristics. For instance, someone might feel like they are "original," but they may not consider themselves "sophisticated in art, music, or literature." Perhaps this trait is hard to measure because the concepts that make up openness—like innovation and originality—can mean a lot of things to different people.

For psychologists, openness in this context refers to having an active imagination, preferring variety, and being intellectually curious. Those things are often prized in our society.

But being "closed" is not a bad thing—it's just different. People who are on the closed end of the spectrum simply prefer *familiar routines* to new experiences. They're more traditional in terms of how they behave and what they think of the world.

So, how does this attribute impact social media use?

People who are open to experience certainly have advantages. They may be more likely to dive into that new feature that a platform offers, which may in turn help them use social media in a more empowering way. For example, when various platforms rolled out the ability to post to specific smaller groups of contacts, it enabled more focused posting with less risk of miscommunications and gaffes. All of a sudden, you had an easy way of posting *only* to political allies about how you thought a political debate went—without infuriating a bunch of friends of friends and maybe losing some of your own friends in the process. People who are more open to experience probably tried out these features first and used them to their advantage.

THE CLOSED ADVANTAGE

It might seem counterintuitive that being *less* inventive or original—in a rapidly changing world like social media—can have advantages. But indeed, the ability to draw boundaries and stick to a known routine has its benefits.

For instance, people who are closed may be more likely to use a healthy and more carefully curated portfolio of platforms. If you're more open, you probably have more comfort with trying out different platforms. This *can* be useful if you happen to find an unknown gem that fits you perfectly. Yet this can also come with a major challenge. As chapter 11 discusses, limiting the number of platforms you use can be better for you, and trying to experience everything can lead to being spread too thin.

Another reason that social media can be hard for people who are unconventional is that it often most naturally plays to the broadest demographic. If a post might be viewed by hundreds or even thousands of people, most people are probably going to err on the side of tailoring that message to the masses. This is why we see a lot of messages about things that are widely known and discussed—like sports, weather, current events, and home decor—but not as many messages about specific, quirkier interests someone might have—like lizard morphology, ziggurat architecture, particle physics, or obscure literary terms.

Then, if someone were to post that message about ziggurats or zeugmas (an obscure literary term), they may stick out in an awkward way.

On social media, being less conventional can lead to the opposite of sticking out—being *marginalized* or *ignored*. As people sift through their feeds at breakneck speed, they tend to positively respond to those things that are familiar, brief, and easy to understand. This is why baby pictures, anniversary announcements, and cute animal photos rack up positive responses. A less conventional message, though, is more commonly skipped over. If it contains a word or idea that isn't familiar, that may be an immediate buzzkill for social media surfers. If a post is too long or involved, that will add to the likelihood that it's overlooked.

Fortunately, because some of the pros and cons of being open or closed are clear, there are some useful ways forward for both groups.

If you are on the closed end of the spectrum, leverage the fact that you're probably already doing a lot of things right. Sticking with the comfort of your routine and focusing on those few platforms that you are used to—and using them in the ways that are familiar—may be a natural benefit. However, you may want to remind yourself to try new features that might help you feel even more empowered on social media.

Another pearl—especially for people who are more closed—is to stick to a dedicated set of things you post about. The idea is that we're most likely to avoid gaffes or embarrassment if we get into a comfortable routine about what we tend to share. For example, I have one patient who is happy with social media because she only posts things in three categories: (1) events like birthdays of family members, (2) publicly praising something that a friend of hers did with a link, and (3) putting up a meaningful quotation. By committing herself to this list, she doesn't feel tempted to toss in something that she might later regret.

If you are someone with more openness, one thing you may want to do is rethink your platforms. Just because 95 percent of people use the same eleven platforms, that doesn't mean that *you* should focus your energies there. There are entire social media sites dedicated to specific interests that might fit you better (see chapter 32 for more on this).

If you're open and are going to stay on the mainstream platforms, though, leverage your natural tendency to *experiment* with tools that can help you MacGyver your way to the kind of experience you crave and deserve. Because social media tends to play to more conventional messaging, try not to take it personally if your message about a historical curiosity you discovered related to the Great Mosque of Djenné gets overlooked.

Instead of changing yourself to fit into the current social media mode, use your natural inventiveness and originality to create spaces and audiences for yourself.

25

Extroversion: Which Way Do You Turn?

***VERSA* IS LATIN FOR "TURN."** So, re-*verse* is to turn again. To be *vers*-atile is to be able to turn quickly.

Similarly, *extroversion* comes from the Latin for "turned outward." *Introversion*, on the other hand, means "turned inward." Extroversion is the last of the big five personality characteristics.

Being extroverted or introverted is about more than which way you "turn," however. Turning or "facing" is *part* of it. Certainly, extroversion and introversion relate to whether you prefer interacting with others or spending more time introspectively.

But according to personality psychologists, how introverted or extroverted you are may have even more to do with how you get your *energy*.

Even if extroverted people happen to spend much of their day on their own, it is those moments of interaction with others that give them the most energy. That's how they get "filled back up." Introverted people, on the other hand, may be forced to spend a lot of their time—because of work or other reasons—interacting with other people. But being by themselves, alone with their own thoughts, is how they best recharge.

Thus, being introverted is not about crippling shyness or an inability to talk to people, as it's often depicted in the movies. It's about how you get your energy.

In some contexts, this world can be tough on introverts. When a teacher leads a standard classroom discussion, the focus often shifts to and from all the extroverts in the room while the introverts watch. Those introverts can start to feel less integrated into the community and less confident about whatever is being discussed.

So, teachers are trained to help everyone receive similar levels of attention and provide ways to ensure introverts get their voices heard. For example, teachers may instruct the whole classroom to write down questions on note cards so that it's not just the extroverts who stand up and have their questions addressed.

But there's a whole area of scholarship across multiple different fields related to how being introverted can be *beneficial*. Some of these books include Susan Cain's *Quiet: The Power of Introverts in a World That Can't Stop Talking*; Matthew Pollard and Derek Lewis's *The Introvert's Edge: How the Quiet and Shy Can Outsell Anyone*; and other titles like *The Introvert Advantage: How Quiet People Can Thrive in an Extrovert World* and *The Secret Lives of Introverts*.

So, when it comes to social media, which is it—do extroverts reign, or do introverts hold the power?

The answer, not surprisingly, is both. I lean extrovert, and this can be helpful in my role as dean of a college of over five thousand students. It's relatively natural for me to walk into a room, immediately start talking to people, and become genuinely interested in their lives. In public arenas, there can be a strong benefit to naturally feeling more comfortable in a crowded room. In these settings, it's valuable that I happen to gain energy from interacting with many people.

That being said, my introverted colleagues have other advantages in high-level leadership positions. They tend to mark their words with great care, which can reap substantial political benefits. They also display great patience during crises: They carefully consider options, plan, and implement the steps to get problems resolved.

Therefore, it's not a surprise that some of the most successful leaders in history have been introverts. Mark Zuckerberg, the longtime chief executive officer of Facebook, is a well-known introvert. His deputies have been quick to clarify that he's interested in people, even when he seems distant. Bill Gates is also known to be an introvert. In interviews he has addressed directly what he thinks are some of the benefits of being an introvert as a leader. However, he also acknowledges that part of his strategy is to hire extroverts to implement some of his ideas.

Strong political leaders have also been introverts. Most historians believe that Abraham Lincoln was introverted. They mention him being able to carefully consider ideas from multiple points of view, being a great listener, and having a rich internal world focused on his own self-improvement. It's not clear whether Barack Obama is an introvert. While he is clearly very comfortable working a room, people who know him also describe him as taking positive time alone to regain his energy. My guess is that he would probably test on the border of introversion and extroversion. A more classic recent introvert president is Jimmy Carter, whose pastimes were solitary pursuits like reading and woodworking.

In terms of social media use, the first issue to tackle is whether being introverted is related to emotional health issues like loneliness in the first place. We do tend to think of introverted people as lonelier. The stereotype of an introvert is someone who wallows in loneliness on a park bench by themselves instead of being in a crowded bar for happy hour.

However, loneliness is often not really about *being* alone—it's about the *perception* of being alone. Psychologists distinguish between *objective social isolation* and *perceived social isolation*. Objective social isolation is when someone spends less time around people on a moment-to-moment, day-to-day basis. But perceived social isolation is when a person *feels* more isolated, regardless of the actual number of people they are around.

These two things don't always go together. Someone may be constantly surrounded by others but still feel profoundly alone. This is an example of someone who has low objective social isolation but high perceived social isolation. It's also why psychologists equate "loneliness" with *perceived* social isolation and not objective social isolation.

A good example of the difference between objective and perceived isolation appears in Colin Turnbull's 1962 book, *The Forest People: A Study of the Pygmies of the Congo.* At one point, Turnbull describes a young Mbuti man named Kenge dancing alone in a clearing. When Turnbull asks him why he's dancing alone, Kenge is surprised. He responds, "But I'm not dancing alone. I am dancing with the forest, dancing with the moon." Then he continues dancing, clearly not feeling lonely.

What this means is that being an introvert does not automatically mean you are prone to loneliness. In fact, when we studied the interrelationships between introversion, extroversion, social media use, and emotional health, we found that people had *the same tendency to feel lonely* when they used more social media, whether they were introverts or extroverts.

This is likely because introverts and extroverts tend to experience different benefits and drawbacks online. No matter where you land on this spectrum, there are ways to enhance your social media experience.

BEING INTRO-EXTROVERTED AND EXTRO-INTROVERTED

Let's start with extroverts.

My extroverted patients initially love social media. After all, it's a whole new way to work a room. It's a whole new group of friends!

For these people, however, the shine often wears off quickly. They are the ones hobbling back a few weeks later saying that they got into an unnecessary tiff. Sometimes that tiff was with someone who had been a good friend, and sometimes it was with someone they didn't even know previously. They also note that what initially seems like genuine connection soon fizzles into surface pleasantries.

If you are an extrovert, the first step is to recognize your strengths and weaknesses in the context of social media. The strength is "easy entry." If you are very comfortable walking into a room and saying hello to everybody, you will probably be more comfortable discovering and using the amenities of a given platform and forming new relationships. You will probably also be more comfortable immediately becoming a "creator" on social media (for more, see chapter 26).

However, you might have more difficulty making social media work for you *in the long run.*

Putting yourself out there in too many places and in too many ways brings risks. A wise extrovert should temper their enthusiasm with appropriate caution and not impulsively jump into all the possibilities at once.

That caution may involve trimming the number of platforms you use. It also might involve focusing on *specific features* of the different platforms. Because I lean extrovert, when I enter a new world, I want to do everything that's available. Only over time have I learned to balance that enthusiasm with more focus.

A final piece of advice for extroverts: You may want to focus on *specific individuals*. This is the kind of thing that comes more naturally to introverts, who usually prefer to form and deepen a few key relationships rather than try to be everything to everybody. There are benefits to both strategies, and optimizing your approach depends on your goals. If you are seeking political office, you probably do want to form and sustain a lot of connections. But my experience with patients suggests that extroverts who can sustain good social media relationships are the ones who focus on a smaller cohort than they would normally be drawn to.

What about introverts?

Introverts sometimes aren't prepared for what they are getting into when they dive into a new social media world. This is because they are coming from an external milieu where they know all too well the challenge of putting themselves out there. So, at first, the online world may seem like a wonderful respite. *Wow,* they think, *maybe this is the opportunity I've been looking for—deep social connection but without that negative visceral sensation I get when I'm physically walking into a crowded room.*

However, many introverts don't recognize how similar offline and online life can feel. After initially feeling more comfortable online, they might be taken by surprise when a familiar feeling of overstimulation and wanting to hide creeps in.

Maybe it's not physically loud and congested in that social media space, but it can still *feel* that way when there is constant chatter from all sides. Maybe there aren't strangers actually looking at you, but it can still seem that way. Maybe most of the time people don't approach you directly, and you can simply watch a conversation unfold. But people might still message you personally about that conversation, or worse, someone might address a comment to you in front of everyone. Now that conversation is doubly awkward: You've got the challenge of communication with that particular individual, but you've also got the discomfort of what everyone else thinks.

While many elements of social media can be a challenge for introverts, they also have natural positive habits that can help them in the long run.

As I've mentioned, introverts generally like to focus on strengthening a few core relationships rather than being friends with everyone. If you are an introvert, a good piece of advice is to *be true to this nature.*

It may be challenging to implement at times. Platforms want users to have interactions with a maximum number of people; platforms don't want users to visit briefly, check in with a few key people, and then leave. Yet, if you can stay true to your natural inclination to focus small, that will likely serve you well.

Here is one way to summarize these pros and cons: Extroverts might have an easier start but more difficulty sustaining their involvement. Introverts can experience the opposite.

In Virginia Woolf's masterpiece *A Room of One's Own*, she writes, "It would be well to test what one meant by man-womanly, and conversely by woman-manly." In a similar way, introverts and extroverts have a lot to offer each other in how they approach different phases of using social media. There might be advantages to being "extro-introverted" and "intro-extroverted." By carefully thinking about both lenses, we can improve how we use social media to enhance our lives.

26
Program or Be Programmed

OFFICIALLY, THERE'S NO SUCH THING as addiction to social media or the internet. The current manual psychiatrists use doesn't have a category for that.

But the rest of us know better, because we've seen this type of addiction—if not in ourselves, at least in others.

There are certain things that define addictions to substances such as alcohol, tobacco, and opioids. Some classic signs of addiction are withdrawal, tolerance, and using something despite the harm it causes.

Withdrawal means experiencing reactions when you stop or having difficulty stopping. How many times have you closed your device and felt an itch to reopen it a few moments later? That is withdrawal, and most people have experienced this with digital technology to different degrees. The urge to keep "using" may not involve physical symptoms, like a racing heart and sweating. But some people do have these physical reactions after ending a session on social media.

In extreme cases, youth have become violent when others attempted to stop their media use. In 2019, an eleven-year-old Indiana boy shot his father in the back while he was sleeping because his father had taken his video games away. In another incident, a sixteen-year-old boy shot both of his parents multiple times—killing his mother—when they took away a particular violent video game from him. After trying to frame his father for the shooting, the boy fled the scene—making sure to take the confiscated copy of the game with him.

What about tolerance? This is the desire to do something more and more as time goes on because it doesn't provide the same rush, excitement, or pleasure that it used to. Something akin to this can happen with social media. At first it is a new and exciting world, and perhaps thirty minutes of perusing stories and feeds leaves a person satisfied. But soon that turns to sixty minutes, ninety minutes, and two hours, yet the whole experience may not be as rewarding anymore.

What about continuing to use something despite the harm it causes?

This is probably the most obvious "addictive" manifestation of social media use. Even if we are doing poorly in school or at work, we often can't pry ourselves away from social media sites—or other digital experiences like video games or the news.

In fact, many people cope with problems at school or work by retreating into what they describe as the "comfort" of digital and social media. A common theme I hear from patients who are having difficulties in their real lives is that sometimes the only comfort they can find is when they pour themselves into dozens of uninterrupted hours of bingeing multiple seasons of a TV show on a digital streaming service. But usually they don't emerge from those marathons feeling better. In fact, they feel guilty that they've neglected other areas of their lives and missed out on more constructive ways of dealing with their challenges.

These dynamics suggest addictive tendencies. But since research into the impacts of digital media is still new—and because things like tolerance are more difficult to measure for social media than they are for chemicals like nicotine—specialists shy away from using terms like *addiction* for social media. Instead they use phrases like "problematic internet use." Nevertheless, professionals recognize that people can exhibit addictive-like tendencies related to social media, like continuing to use social media even if it is causing financial, occupational, or social problems.

As I've discussed, designers go to great lengths to get people "stuck" to social media platforms, so it shouldn't be a surprise that, at the extreme, some people may exhibit addictive behavior with social media.

There is an established scale used to measure "problematic social media use." It asks commonsense questions to get at whether people have exhibited addictive qualities around social media. For example, it asks people how much distress they would feel if their social media were taken away, if they have ever felt the urge to use more and more social media, and if they tend to use social media despite any harm it might be causing them. Although the scale was developed specifically around Facebook, my research group has had success in using it for social media in general.

This led to an insightful study that my colleague Ariel Shensa and I completed. What if two people both use the exact same amount of social media—let's say two hours per day—and they are also matched in terms of things like sex, race, relationship status, and living situation. In fact, the *only* difference is that one admits to addictive tendencies on social media and the other doesn't.

Do these two people—matched in every way except for being flagged on the problematic social media use scale—have similar risks of being depressed?

The results of this study were remarkable. Among a national sample of young adults, the people who acknowledged addictive tendencies around social media were more than *three times* as likely to be depressed. This means, in addition to considering the amount of time we spend on social media, we need to consider whether we are experiencing *addictive tendencies* while using it. This may be just as or even more important to our overall well-being.

In his 2010 book *Program or Be Programmed*, scholar and media theorist Douglas Rushkoff says that there isn't much point in arguing about whether the internet is good or bad for society, because it's clearly part of our lives now. Instead, the choice we have to make is whether we are going to be *controlled* by the internet—which, as we've seen, often tries to "program" our behavior—or whether we are going to *control* our digital interactions and become the "programmer."

This includes social media. If we only passively consume social media, we risk being "programmed" to the point of addictive behavior. We risk not only our time and attention but also our emotional health. Rushkoff's antidote is to encourage active participation, so that we become the "programmer" of our own digital world. This means we should intentionally choose what we click on, how we spend our time online, and what we internalize as we scroll the internet—because we are what we click. Active decision-making and participation in social media is how we empower ourselves on platforms designed to manipulate us. Hence the importance of being *creative.*

Hard data suggest that being intentional and actively making choices online is good for our emotional health. Psychiatrist and researcher César Escobar-Viera and I conducted a recent study with about seven hundred young adults. Using a five-point scale, we measured how "passive" or "active" each user was on social media. Passive users were those who mostly just scrolled through a feed, while active users were those who tended to control and create content.

Even when we controlled for demographics, every one-point increase in passive social media use was associated with a 33-percent *increase* in depression, while every one-point increase in active social media use was associated with a 15-percent *decrease* in depression. Being more passive online went along with more depression, but more active use was associated with less depression.

HOST AN APPY HOUR

To fight off any addictive urges with technology, we need to program instead of being programmed. We need to take control of our digital environments. We need to create and not just consume. But how exactly do we turn ourselves into programmers?

In order to take command of a platform or app, we first need to deeply understand our digital environments. It's surprising how many of us use technologies, like social media platforms, but don't know how to do something quite basic—like turn off our notifications. (In our defense, platforms make it very hard to manage notifications.)

One way to actively get to know a platform is to have a dedicated digital media "party." Call it an "appy hour." The purpose is to get together with other people and help one another figure out the tips and tricks specific to the apps and platforms you use. You can even create specific tasks and introduce competitions. Who can turn off all notifications first—go! As goofy as it sounds, this can be fun. And it's just one of the many ways you can learn the full scope of a platform and the apps you use.

Another way to develop a deeper understanding of your digital life is to take the phrase "program or be programmed" quite literally: Learn to program computers, even in a small way.

When my kids were young, just like all young kids, they wanted devices. Cell phones, laptops, you name it. Therefore, I introduced a simple household principle—learn to program something on that device, and you can have it.

Maybe that sounds overzealous. My kids at the time were both in elementary school, but they followed through. Before they could have their own devices, they used a family computer to visit Scratch (scratch.mit.edu)—a digital programming environment for kids created by professionals at the Massachusetts Institute of Technology—and they created their own games. Not surprisingly, they didn't just create the one game necessary to meet my demand. They kept programming.

When it was time for a laptop, my kids learned HTML and CSS. There are easy, free tutorials online. Now both of my kids have their own personal launch pages that they coded using real computer languages—and they can change the content on those launch pages whenever they want.

Neither one of my kids is interested in pursuing computer science as a career, and that wasn't my intention. I wanted them to become empowered. Now, when they see a web page, they have an idea of how it was created. It's sort of like working in a restaurant. Afterward, you know what happens behind the scenes, and so you know what to order—and maybe what not to order.

Choose to be empowered, whatever that means for you—actually learn Python if you want, or just have a party with friends to explore how your apps function. But choose *something*. It will help you feel that much more comfortable with and in control of your digital experience.

27
Develop Social Media Literacy

MEDIA EDUCATORS COMMONLY PLAY "name that brand" with their students. They flash logos on a screen, and without any prompting, people light up with recognition that the large R is from the Houston Rockets' logo, the S is from Starburst, or the T is from T-Mobile. Students lean together, point excitedly, and enjoy the game.

But I suspect that there's more to their smiles than just that they guessed correctly. They are happy because they are responding to what the advertising world calls an *emotional transfer*.

To create an emotional transfer, an image is connected to an emotion. In the case of a Coca-Cola advertisement, the ultimate aim is usually to convey happiness. Coca-Cola does this deliberately and explicitly, using taglines like "Open happiness" and "Have a Coke and a smile." Yes, the ultimate goal is to sell a liquid—but they do that by selling the idea of happiness.

The emotional transfer occurs when an ad evokes an emotion in viewers, who then transfer it onto the product. For example, the first twenty-seven seconds of a thirty-second commercial might tell a heartwarming story filled with nostalgia and warmth without even mentioning the product. Then, at the end of the emotionally charged story, the logo for a particular toaster will slowly fade in. Like Pavlov's dogs, we now connect that toaster with nostalgia and warmth, and the more times we see that advertisement, the more powerful that connection becomes.

Understanding these kinds of processes is what *media literacy* is all about. A common definition of media literacy is "the ability to access, analyze, evaluate, and produce media in a variety of forms."

Every consumer of media develops a certain amount of media literacy. But it takes conscious effort and intention to more deeply recognize strategies like emotional transfer and to learn to resist them. Companies and marketers want their strategies to work subconsciously. But even when we know what's happening, these approaches can still work. With enough repetition, if we subconsciously connect McDonald's with happiness, then we're more likely to pull into the drive-through, because we connect those fries with experiencing happiness. For McDonald's, "I'm lovin' it" and "Smile" have been two of their most successful taglines.

Developing media literacy may help us be less affected by marketing. For instance, do you know why the colors red and yellow are used in many logos and advertisements for food and drink? It is because experiments have shown that humans connect yellow and red with hunger and thirst.

Media literacy is about gathering information and increasing our awareness so that we can *critically assess* the marketing we see. For instance, McDonald's ads often focus on happy children, and indeed, McDonald's has been the largest distributor of toys in the world via its "Happy Meal"—but its food has also been a contributor to the epidemic of childhood obesity. Similarly, fast-food chains often emphasize happy employees, but in truth, how happy are people in these low-wage jobs, where an employee might take months to earn what the CEO makes in an hour? Thinking about these things—and noticing the irony between what ads promise and the reality—might make us less likely to purchase that company's product.

One way that media literacy works is by bringing the subconscious into the conscious. One Harvard professor suggested that as much as

95 percent of advertising works subconsciously. Consider the Amazon logo. You've probably seen it thousands of times, but have you ever noticed that the arrow connects the A and the Z in the word *Amazon*, which represents how the company "has everything from A to Z"? You probably noticed that the arrow also doubles as a smile, but did you ever consider why that specific font was chosen? Or why all the letters are lowercase and sans serif? All these things were carefully decided to maximize its impact on you.

Media literacy has become prominent in educational circles. It's now recommended in school curricula to help kids learn about language arts and social studies. It's suggested by the American Academy of Pediatrics to help kids be less affected by advertisements for things like smoking, vaping, and alcohol use. It's recommended by arts organizations to help people understand the artistic process and appreciate the arts more.

But applying media literacy to *social media* hasn't been sufficiently explored.

While social media isn't exactly advertising, there is a lot of advertising *on* these platforms. Depending on the site, about every third to fifth post may be an advertisement. Advertisements often line the sides and headers of each page. This is one reason why we need to be proficient at analyzing and evaluating what we see on social media platforms.

Even more than that, media literacy can help us analyze *all* communications, including the actual content of social media posts. As I mention in chapter 4, there are reasons to believe that social media may be even more influential than advertising when it comes to our emotional lives.

The reason for this influence is that social media *combines the power of production values with the influence of peer-to-peer communication.* A magazine advertisement for tequila is a work of art: It's a carefully crafted image that involved any army of executives, psychologists, graphic designers, photographers, and models—not to mention huge amounts of cash.

But even if the advertisement works—and we subconsciously begin to link this tequila with concepts like fun, camaraderie, and social acceptance—we still recognize that the ad does not represent "reality." We realize that the positive image was made to sell a product.

The world of social media, however, is different. Each photo has not been doctored by a team of graphic designers. But each photo *has* been selected from hundreds of possibilities. Each word accompanying the post has been selected extremely carefully by the author to convey a specific message. In this way, the social media we are experiencing has been heavily curated.

Then that extreme curation is paired with the power of peer-to-peer communication. A lot about a social media experience feels "real." After all, it's the product of real individuals discussing real events.

Thus, on social media, it's even more important to consciously evaluate and critically assess what we're seeing and reading.

THE AD IT UP PARADIGM

I developed a mnemonic to make it easier to analyze media messages: AD IT UP. It's based on the basic principles of media literacy. We've used it to create programs that have helped adults and kids analyze all kinds of media messages, including tobacco advertisements, violence in movies, food commercials, and health-related scenes in television shows. Each letter refers to a question to ask ourselves as we look at a message. By answering those questions, we develop and engage our media literacy. Here is what the acronym stands for along with the questions:

Author: Who is the author of this message, and why did they make it?

Directed (audience): Who is the message directed at? In other words, who's the intended audience?

Ideas: What kinds of ideas do they want to get across?

Techniques: What techniques do they use to get across those ideas?

Unspoken: What's unspoken or left out of the message?

Plan: What's my plan now that I've gone through this process? Does this change my thinking or behavior?

You could use this simple system to think more deeply about a single social media post. For example, if someone's rosy post leaves you feeling jealous, the U in "unspoken" might remind you that a lot is *left out* of that person's message. Not every day of their life is as rosy as this one, and that realization might temper your feelings.

The experience of social media is more than single posts. This system can also help you analyze the entire set of communications you're fed. Social media platforms brilliantly tailor their features and advertisements to you. This is because they have so much data about you. This includes the factoids you entered when you joined, the pattern of what you "like" and respond to, the characteristics and pastimes of the people you respond to the most, and your search history.

When you see advertisements or promotions, think of the letter D and ask: Why is this message *directed* at me? This can help you think consciously about your vulnerability to that kind of advertisement. Then consider T, or the *techniques* used to sway you, like emotional appeals, colors, fonts, logos, and lighting.

The AD IT UP process culminates with P—thinking about your plan for *responding* to all this information. For example, after examining in detail the various reasons that social media makes it seem like everyone else is having a rosier life than you are, you might commit yourself to thinking more *horizontally* rather than *vertically* (see "Think Horizontally," page 31).

As chapter 4 explores, social comparison is baked into us—like other animals, we tend to instinctively compare ourselves in terms of a "vertical" hierarchy. By thinking more horizontally, we choose to avoid only considering who's doing better and who's doing worse. Instead, we appreciate the fact that everyone has different situations, challenges, and successes. This simple shift of perspective works for many people. When someone shares their big success on social media, instead of feeling knocked down a peg on the hierarchy, we can feel good for that person and how they're faring in their unique circumstances.

In this way, actively answering these AD IT UP questions can help you to develop actionable plans. Ads and other media messages can become your friends in growth rather than things to avoid.

Ultimately, cultivating our media literacy never ends, but it involves the same approach: to actively analyze, evaluate, and understand what is behind the media messages we receive. In this way, we become active participants in the communication process. Then, once we are actively interpreting communications instead of just letting them wash over us, we're less likely to be simply dragged along for the ride—and more likely to be the one driving.

28 Consider Your VQ

OUR SOCIETY TENDS TO FOCUS on one type of intelligence, mathematical-verbal, which is measured by tests like the IQ and the ACT. Scores on those tests can correlate with measures of life success. But studies show that a person's "EQ"—or "emotional quotient"—also correlates with key life outcomes. And, in fact, EQ can in certain contexts be even more important than IQ.

When it comes to social media, it's important to consider what I call our VQ—or "visual quotient." This refers to how much *visual* material we're exposed to online. While this is not a kind of intelligence, it can be thought of as a "quotient." It involves asking questions like these: What kind of visual elements are you surrounding yourself with? How often? And how are those visual elements affecting you?

On social media, thinking about our VQ might be most critical when it comes to issues involving body image and physical attractiveness.

On the TV show *The Good Place*, the character of Tahani Al-Jamil is played by the actress Jameela Jamil. While the character Tahani is a lovable if condescending name-dropper who likes to pretend she's an activist, Jameela herself is an actual activist in real life. One key issue she's concerned about is the portrayal of weight on social media. Jameela feels that celebrities who flaunt impossible weight ideals and sell diet pills and supplements offer false hope and promote toxic notions about body image. She wants to help young women boost their self-esteem regardless of the number on the scale.

Unrealistic body ideals can influence eating habits and lead to eating disorders. As a clinician, I've seen the massive harm these conditions can cause. Studies show that those diagnosed with anorexia nervosa have a death rate of about 10 percent. This is extremely high. Compare that with stage 1 breast and prostate cancer, which kill only about 2 percent of patients in the first five years.

While genetics play a role, eating disorders are also deeply related to environment and society. In a classic study conducted in Fiji in the mid-1990s, anthropologist and psychiatrist Anne Becker decided to test the relationship between body image and media. Knowing that television was about to be introduced to Fiji, she surveyed girls before and after the arrival of TV.

Before television, the comment "you've gained weight" was considered a compliment in Fiji, and body expectations reflected family and social values. After television, Becker found that Fijian society adopted a "thinner is better" mentality. Girls at risk for eating disorders jumped from 13 percent to 29 percent. The proportion of girls who vomited to control their weight quintupled from 3 percent to 15 percent. Becker theorized that this change was caused by increased exposure to impossible body-image ideals on television, where beauty is typically represented by thin models and actresses.

Today, social media images often embody this same "thinner is better" ideal, and it may have a similar effect. For obvious reasons, when people post images of themselves, they typically share only the most attractive photos. Yet many people also manipulate these images so they seem thinner than they actually are or project an idealized version of beauty.

Of course, no one wants to post unflattering images. For this reason, it's important to remember that social media, just like TV, does not present a realistic portrait of healthy people of all shapes and sizes—and in all forms of unfiltered beauty. This is another way that social media, though it seems to represent "real life," is in fact not very realistic at all.

CAN SOCIAL MEDIA AND POSITIVE BODY IMAGE COEXIST?

Concerned about the effect of social media on eating concerns and disorders, my colleague and public health researcher Jaime Sidani and I conducted a study among eighteen hundred young adults. We asked participants to agree or disagree with statements like "Someone (such as a health professional, a family member, or a friend) has expressed concerns about my eating patterns," "My weight negatively affects the way I feel about myself," and "Food dominates my life."

What we found was that increased social media use was strongly associated with more eating concerns. Compared to a group who used the least social media, those who used the highest amount had two to three times the risk for eating concerns.

So, how can we use social media *and* protect our body image?

This is a critical question. Until the day comes when people feel comfortable posting unfiltered, raw photos of themselves even if they don't look their best, it will be up to us as viewers to change how we consume content.

This begins by being more conscious of the messages we are receiving from images. By itself, it usually isn't enough to simply tell ourselves *Yes, I know everyone looks thin and gorgeous in their pictures and profiles, but that isn't reality*. However, reminding ourselves to be self-aware—to make the subconscious conscious—may help create some observational distance and lessen the effects.

The other thing we can do is actively think about our VQ as we make decisions about which platforms to use. For example, some people prefer Reddit—and other more heavily text-based platforms—because they can be less visual.

However, all social media platforms are becoming more visual. Originally, Twitter was mostly a bunch of text scrolling down a page.

Now there's nearly always a picture with each post. Instagram is by definition all about images. A few studies have ranked social media sites in terms of which ones are linked with more self-doubt, and they raise the most concern around Instagram—probably because of its emphasis on visuals. Nevertheless, it remains massively popular. Instagram launched in October 2010, and a decade later over a billion people used it every month.

Another thing to keep in mind are your "big five" personality traits and how they affect your social media use and experience (which is explored in chapters 21 to 25). We all react to social media differently based on our personalities. Take the time to honestly assess your VQ as it relates to body image, weight, standards of beauty, and attractiveness. If you notice that visuals on social media bring down your self-esteem or mood, make an effort to change your VQ.

Body-image issues did not arise because of social media. Instead, social media reflects the society, media, and culture we live in. Social media might seem like a Pandora's box of problems, but we're not powerless. As with any issue, the first step is awareness of the problem and actively managing our expectations.

Most of all, though, we must hone our responses by knowing ourselves, thinking consciously of how visuals affect us, and crafting our social media environments in response. Maybe this means unfollowing particular people. Maybe it means switching platforms. Maybe it means taking steps to make ourselves more aware of how common filters are and how dramatically they can do just that—*filter* reality.

These techniques won't take away all the risk, but they represent creative ways to decrease the challenges and increase the value we get out of social media.

29

The Eloquence of Emoji 😊

EVERY FALL, THE *OXFORD ENGLISH DICTIONARY* presents their "word of the year." In 2018, it was *toxic*. In 2016, it was *post-truth*. In 2013, it was *selfie*.

As with *selfie*, sometimes the word is new to the dictionary because of a new technology, event, or idea.

Toxic, on the other hand, represented an old word with new meanings. In the past, *toxic* had been mostly reserved for chemicals. But in 2018 people increasingly used metaphoric phrases like "toxic masculinity," "toxic relationships," and "toxic culture." Prominent news stories related to "toxic algae," "toxic gas," and "toxic waste" also contributed to the rise of *toxic*.

In 2015, something interesting happened. For the first time in history, the *Oxford English Dictionary*'s word of the year *wasn't a word*. It was this: 😂

As you probably know, this is an *emoji*—a pictograph that was originally named because it conveys some type of *emotion*. Today, however, the term refers to any type of similar pictograph regardless of emotional content. You can find emoji (the plural can be *emoji* or *emojis*) of an ice skate (⛸), a pineapple (🍍), a swirl of poop (💩), or just about anything else.

There were mixed reactions to the news that the *Oxford English Dictionary*'s word of the year was an emoji. Many linguistic purists were upset. Did this signal the beginning of the end of language? Ultimately, will we be reduced to a society in which we just use pictographs to describe all the complex things our brains are capable of thinking and feeling? (Interestingly, pictographs were a major part of how language *began* thousands of years ago, but I digress.)

I'm not so sure that the emoji marks the end of civilization, and I think it's probably better to understand, embrace, and express our creativity with emoji instead of railing against them.

Long ago, when words were artfully scripted using quills, an innovation enabled mass production of texts. However, opponents of this technology said that it wasn't a good idea for texts to be "spread about like dung" to the world. They said that this technology must be the work of Satan because it seemed magical and destined to take artistry out of scripting. Typographers were charged as witches.

That innovation was the printing press.

Considering the advances the printing press enabled, we need to be cautious not to automatically reject new technologies because of our fear of the technology itself or fear of change.

Let's consider the particular emoji honored by the *Oxford English Dictionary*—which is called "Face with tears of joy" (😂). It basically means that you are laughing so much that you're in tears. This is not to be confused with "Rolling on the floor laughing" (🤣), "Loudly crying face" (😭), or "Grinning face with sweat" (😅). Why was it picked in the first place?

One reason was its popularity. In 2015, it was estimated that it made up 20 percent of all emoji used in the United Kingdom and 17 percent of all emoji in the United States.

It's also an expressive and meaningful image. It can completely change the gist of whatever text it's paired with. For instance, if someone posts an image of an actress and writes, "I love her new hairstyle," most people would assume that, well, the person loves her hairstyle. But "I love her new hairstyle 😂 😂" means something totally different.

Before emoji became commonplace, there were "emoticons," which are combinations of letters and other characters to make "faces" that convey emotion. For instance, XD resembles a sideways face with eyes squeezed shut and an open, howling mouth. Similarly, ;'-) and :-(are other common emoticons representing sideways happy and sad faces, respectively.

But let's face it: 😂 is basically the best way to convey 😂, short of seeing someone actually guffawing so much that they're teary.

There's also something endearing, efficient, and effective about conveying an emotion in a single, simple image. When someone posts about going through a hard time and you respond simply 💔 or 😔, this conveys a great deal. A few simple curved lines show solidarity in sadness.

In other words, emoji can be a powerful tool in your communication toolbox. So use them freely, and don't worry about naysayers and critics. You aren't going to forget how to write or talk.

Do you need some ideas for emoji? Here's a secret weapon you might like: Emojitracker.com. This site, in real time, shows you every emoji that is being posted on Twitter. Each emoji is followed by the total number of times it's been used, and each emoji and its number flashes green at the moment it's being used. For example, see figure 5. At the time this chart was captured in 2021, 😂 had been used 3,127,480,709 times. For some perspective, imagine that 😂 was being used once per second. In that case, it would take about a hundred years for it to be used 3,127,480,709 times.

It's interesting to use the emoji tracker to see which emoji are going in and out of fashion. It's also a visceral way of experiencing just how central emoji have become to our everyday communication. As green flashes wildly before you, you can't help but think of all the people sending these messages across the world.

3127480709	1563954793	1116434416	1110243387
968658041	803048120	707630662	544152791
528482628	524795439	492733009	472440129
430251310	425205711	389872072	368728513
360405901	338257616	326544315	320682566
314758391	295914645	286839683	286047045
276738175	276257911	273934801	269112696
257550879	255773873	238701333	234986819
222048006	214929348	209543101	203136341

FIGURE 5: Emoji Traffic on Twitter

The emoji we use also tell us something about ourselves as humans. In figure 5, there are plenty of sad images; however, emoji like these are often being used to *acknowledge* someone's grief and to help *share* it. And the most common sentiments we see are those of laughter, love, and positivity. So keep using emoji creatively. They just might help you "say" the right thing at the right time.

30

Become an Alert Wizard

IN 1961, KURT VONNEGUT PUBLISHED a short story called "Harrison Bergeron." It portrays a dystopian future United States in which citizens are not allowed to be more athletic, attractive, or smarter than others. This "equality" is achieved by introducing *handicaps*. A graceful dancer has to wear heavy weights. Attractive people have to wear masks.

What about people who are more intelligent? Well, they are exposed to random noises and interruptions that disrupt their attention, concentration, and thoughts.

At the time, this was a dystopian vision—yet this is more or less the situation in the early twenty-first century *for all of us*. "Notifications" regularly accompany us all day, every day. Sounds, vibrations, and pop-ups from our devices distract us from giving anyone or anything our full focus.

In "Harrison Bergeron," a similar interruption leads to one of the central tragedies of the story. An intelligent man, because of a well-timed interruption, doesn't even realize that his son has been seized from his family by the government.

Hopefully, our interruptions are not as dramatic or as tragic. But what is the opportunity cost of our notifications? How much more effective, creative, and peaceful might we be if we were not constantly interrupted in this way?

We can be significantly affected by our smartphone *even if it's off.* A 2014 study published in the journal *Social Psychology* involved two groups of young adults who were asked to complete certain tasks. One group was instructed to have their cell phones—*turned off*—placed on the table beside them. The other group completed the same tasks but did not have their cell phones on the table.

For extremely simple tasks, there was no difference between the groups and their ability to function. However, when the tasks were more complex, people without cell phones nearby performed significantly better compared with those who did have cell phones near them. Just the *presence* of the phone—even without any beeping or buzzing—was enough of a distraction to impair people.

Why did this happen? The researchers surmise that we've been conditioned by our phones to *expect* interruptions. Even if it's not active, just the existence of our phone triggers the subconscious feeling that at any time there may be an alert, call, or notification. This saps our mental energy and focus. Our phones also act as subconscious reminders of things we need to do. They trigger our desire to make sure there isn't a critical email or text we simply must act on *now*.

If this is what a phone can do when it is off, imagine how significantly impaired we are when it is on and literally calling for our attention.

ATTENTION IS OUR MOST PRECIOUS COMMODITY

In today's nonstop world, it can be argued that our attention is our most precious commodity. But many of us give it up every day—minute by minute—to maintain the feeling that we aren't missing anything.

Why don't we take more control? Why don't we make determined, conscious decisions to simply click the right buttons and take back our attention? *Why don't we turn off our notifications?*

There are three major barriers. The first is that we are often ambivalent about turning off notifications. The second barrier is that it is not a simple task to turn them off. And the third is that turning off notifications is not a quick fix—it's an ongoing issue. Let's explore how we can overcome each of these obstacles.

Take Back Your Time

Most people recognize that FOMO (fear of missing out) is more of an anxiety than a real concern. But it's still hard to turn off the phone because people often feel that they need to remain available to others. What if something important comes up? What if people need me? What if there's an *emergency*?

This is something only you can decide for yourself. Some people need to leave notifications on for their work or their family. In general, though, we overestimate the need to be connected at all times. After all, what did people do twenty years ago? Somehow, we survived. Are we so much better off today because we are notified of every post containing a lovely picture of an acquaintance's lunch?

We need to strike a healthy balance. We aren't stuck between being connected and being stranded on a desert island. The truth is that there is a lot of middle ground.

The irony of the situation is that, in our fear of not being connected, we *lose* connections. For example, if we are with someone and our phones are constantly grabbing our attention, we're impairing our actual connection in the moment because of our fear of being "disconnected." We need to be able to give up *some* connectivity in order to step back and enjoy our lives more.

In my position as dean of a large college, I am responsible for over five thousand undergraduate and graduate students and hundreds of faculty and staff members. Urgent issues arise frequently. As a result, we have strong systems in place so that I and my executive team can be notified immediately if there is an urgent issue that requires our attention. But this does not include notifications on my devices. That would distract me from sitting with a faculty member or student and listening to them carefully. It would keep me from efficiently working on proposals and projects related to improving people's lives. And it would keep me from optimally dealing with *actual* urgent situations.

Make a clear choice for yourself about which parts of your life need more of your undivided focus, and choose which notifications you want on and which ones you don't. Then turn off your phone and place it in another room so you can be present for the things that matter to you.

Turn Off Notifications (Even If You Need a Ten-Year-Old's Help)

Once you've decided which notifications to deactivate, all that's left to do is simply turn them off, right? If only it were that easy.

The default for most devices you own and apps you install is for notifications to be *on*. It's not in the financial interests of social media platforms to make it easy to turn off notifications. Often the selection is buried within a menu within another menu and behind an icon of questionable intuitive value. In my case, it required not only concerted effort but also the occasional assistance of my then-ten-year-old daughter for IT support.

Use whatever resources you can muster. Take the time to get to know every app or platform you sign up for and understand where all the extra noise is coming from. Then, go through each and turn off anything that isn't essential.

Think Long Term

Finally, it's important to realize that this is an ongoing issue. The average person has between sixty and ninety apps installed on their phone. They launch about ten each day, but others are often operating in the background, and many provide their own alerts. And these sixty to ninety apps are constantly cycling as we excitedly download the next one. If we commit to turning off notifications and taking back some of our scattered attention, it necessitates an ongoing commitment. It's not something we can do once and be done with.

There are two things you can do immediately. The first is to make it a habit to immediately turn off notifications for every new app you install. It's best to do this *when it's installed* because otherwise your likelihood of ever getting to it goes down a lot. Also, when you download something, you're usually the most excited about using it, so do this while you're checking out everything else about that app. The second thing you can do is to delete at least one app (preferably more) whenever you download a new one. Aside from reducing unwanted notifications, this will also clear up space on your device and reduce the likelihood that apps will fight with one another, negatively affecting performance.

We need to think carefully and conscientiously about our devices and choose how they interact with our actual lives. Read "Harrison Bergeron" for some motivation. Have coffee with a friend and discuss these issues—or even host an "appy hour" (see "Host an Appy Hour," page 159).

But that's just the beginning. Then comes the process of actually making the changes and sustaining them over time. It's not that hard, though there's a learning curve as you get the hang of notifications—and other aspects of your apps.

And it can't hurt to have a tech-savvy ten-year-old close by.

31

The Good Stuff Is Often in the Back of the Room

LET'S SAY YOU WALK INTO a vast social hall for a dinner party. There are stations with different foods set up all over the room. You know that there will be a station in the back with your favorite food, lentil soup. In a different corner of the back of the room, there will be black bean tamales, your second favorite food.

But as you walk in the door, several servers run up to you with plates of macaroni salad and marshmallow ambrosia. You are tempted, perhaps, but these are not the foods you prefer. So, do you pig out on what's right in front of you, or do you walk to the back and get your favorite foods?

This is not a trick question. The answer is obvious. Of course you keep walking to the back of the room and enjoy what you really like. Unfortunately, however, in the world of social media, we often feast on the macaroni salad and ambrosia right in front of us, get full, and go home.

That is, when we enter our social media feed, we are often seduced to spend our precious time on what I call "low-quality" content that we're not actually interested in. Social media platforms populate your queue based on complex and varied algorithms. But the one thing that these algorithms have in common is that they optimize that feed for their own financial reasons. They make more money if you eat the ambrosia rather than the tamales.

According to the insider documentary *The Social Dilemma*, many social media platforms also line up content for you that contains certain key words or phrases that are more likely to end up selling you something. So, for example, on YouTube we may be preferentially nudged toward content with more embedded advertisements. On Facebook and Instagram, you may be fed the content with the most prior reactions because these posts are more likely to engage you emotionally and keep you on these sites longer.

By simply starting at the top and scrolling down your feed, you are consuming the low-quality stuff that the aggressive servers are offering. Behind them is the high-quality stuff in the back of the room—the people and content you really want to interact with. We need to politely reject the aggressive servers and move beyond them to get what we really care about. We need to intentionally curate the experience we want to have every time we log on to social media.

How, specifically, do we do this? One way is to focus on small groups, including chat clusters with one or a few people at a time. This can help you focus on the people you feel closest to. Those people are more likely to support you and understand you if you have something challenging to talk about.

As chapter 15 discusses, bad experiences can be powerful, and they can easily overshadow our positive experiences online. Communicating with smaller groups substantially reduces the chance for gaffes, misunderstandings, and other related experiences that can make social media sessions go bad.

In addition, when interacting with one person or a few people who know us well, they will likely overlook—or try to understand—anything that is awkward or not clear. Someone we don't know very well may not afford us the same latitude. As we know, social media fails, autocorrect errors, and other miscommunications are common. So, it's a good idea to protect yourself by having the person on the other end be the type of individual who will cut you some slack.

Another important reason for communicating with one or a few people at a time is that it makes the person or people we approach feel more special. One of the great benefits of social media is that it enables us to express gratitude, compassion, and similar sentiments every day and at any moment that's convenient to us. Years ago, when you felt inspired to tell someone you were thinking of them, you had to find a card, a pen, an envelope, the time to write, a stamp, and the wherewithal to get to the mailbox or the post office. Now it just takes a couple of seconds. These are the gifts of transitioning from low-quality to high-quality interactions on social media with smaller groups and one-on-one communication.

CELEBRATE FRIENDS . . . BUT NOT ON THEIR BIRTHDAYS

One concrete way to transition from low-quality to high-quality interactions on social media is to celebrate birthdays differently.

Birthdays on social media have become something of a standardized event. When it's our birthday, we often get a flood of well wishes from many people, including old friends we never hear from at any other time. Some people send animations and emoji; some write a simple greeting; some people write a paragraph; some get wildly creative. It's kind, enjoyable, and heartwarming.

But it's also a routine. People haven't necessarily remembered our birthday; social media has. And some people send canned greetings or do nothing more than click a button that automatically enters a birthday greeting of the platform's choice. It's not that the greeting isn't appreciated, but it's not always very special or particularly thoughtful.

Don't get me wrong—I like birthdays on social media. I enjoy seeing people's creativity. I like seeing the names of the people who wish me well. I'm one of those people who logs on, feels good about all the messages, and writes back to as many people as I can. I certainly do not want to abolish birthdays on social media.

But there's a missed opportunity here. The purpose of messaging someone on their birthday is to connect with them and let them know we care about them. If it happens to be the birthday of an old or distant friend we don't interact with often, shouldn't we make it a *meaningful* interaction?

Rather than—or in addition to—posting on someone's actual birthday, I suggest engaging with friends on a less hectic day. In the world of social media, your heartfelt message on someone's actual birthday can get lost among the countless other birthday wishes. And after receiving so many messages, people don't tend to respond in a thoughtful, individualized way to everyone.

So, pick a *different day* to send a celebratory greeting. One possibility could be three days before someone's birthday. On Facebook, for example, you can visit the "upcoming birthdays" page, which will list everyone you know on Facebook. Since the person's actual birthday is three days away, they are less likely to have 120 other messages to sift through.

This is not the only way to exercise this kind of strategy, of course. It doesn't have to involve a birthday wish at all. Pages like the "list of upcoming birthdays" can simply serve as your *organizational system* for making sure that everyone is on your radar at least once a year. You can also work several months ahead and send an "I'm thinking of you" greeting.

This birthday technique is just another way of getting creative to turn a low-quality experience—a big plate of ambrosia—into a high-quality experience. Keep thinking small, focusing on deeper relationships, and nurturing meaningful interactions, and you'll be on your way to an improved digital life.

32

Love Thy Neighbor, but Choose Thy Neighborhood

THE TOP TEN OR SO social networking platforms account for about 95 percent of all social media activity.

Yet there are hundreds—if not thousands—of social media platforms. This means that the vast majority of people focus on a small proportion of the options out there.

One reason for this focus on a few top sites is that those sites represent big businesses that have the vast majority of marketing dollars. Despite challenges and downturns, they can continue to keep themselves relevant. But sites like Care2—a social activism and green living social network platform—often can't keep up.

People also flock to the top sites because of personal referrals. If your real-life friends start sending you Instagram links, you're more likely to join Instagram, perpetuating the cycle of the growth of the largest networks.

Another reason is the "bigger is better" mentality. Even if someone is not referred by a friend, if they are thinking of seeking out new contacts—or digging up old ones—they'll likely consider a site like Facebook or Instagram to be the most promising. Especially because social media is about connections and exposure, it's understandable that most people will choose communities that have the highest traffic.

Part 2 discusses the importance of being selective about how many platforms you use. It's also important to carefully consider if you've found the *right platforms* that truly nurture your interests and goals. If you haven't, it's time to go beyond the big ten.

This chapter's title is taken from a wise and clever quotation attributed to John Hay, the US secretary of state under William McKinley and Theodore Roosevelt. In today's world, we have not only physical neighborhoods but also digital neighborhoods. We absorb the values of the people and communities we surround ourselves with—whether we're at a coffee shop or in a chat room. So, when considering which "neighborhoods" you want to inhabit, carefully think about questions like these:

How do I want to be treated?
Does this platform support my values?
How do the people on this platform make me feel?

We all want to reduce negative experiences on social media. By selecting a platform that focuses on a narrower range of topics, you may decrease the number of negative experiences you'll have. Honing what you are consuming is also more likely to focus your time online on platforms that support and inspire you.

As an example, DailyStrength is a social networking platform that caters to people with mental health concerns like bipolar disorder, anxiety, and depression. DailyStrength forums discuss the kinds of things that people talk about on other social media sites—movies, music, advice about purchases, news, and so on. The difference is that the users of this platform are engaging with others who have *more sensitivity* toward mental health conditions. That doesn't mean that there are never any insensitive things said. It just means that it tends to be a more supportive and understanding group.

The same kind of thing exists on hundreds of other social network sites, each of which has its own ecosystem. Finding the ecosystem you prefer will help you tailor your social media experiences.

Another example is Ello—a social media platform that is specifically geared toward people working in and interested in the creative arts (its URL is ello.co, not ello.com, which produces dietary supplements). At Ello, you can subscribe to communities discussing and posting about painting, photography, writing, music, fashion, art and minorities, collage, and minimalism. If these are things you are interested in, this community—compared with something like Facebook—will probably be more satisfying, since it is devoted to kindred spirits.

VIRTUAL AND METAPHORIC MONSTERS

As a final example of the kinds of communities that can be found in the little-known corners of the social media world, there's Habitica. It's an open-source project that works somewhat like a role-playing game, such as Dungeons & Dragons. You create a character for whom you can buy armor, weapons, and potions. The difference is that your character advances not based on a roll of the dice but based on how you are doing *with real-life habits* that you are interested in changing or strengthening.

To begin, you make a list of habits that you want to do every day, like "do thirty minutes of yoga" or "eat at least twenty-five grams of fiber." You can also keep your "to do" list of chores, errands, and projects on this platform. Then, as you check off things you do successfully, you get points that translate into more value for your character, including more experience points, better health, and treasure. But beware, because if you don't do the things you are aiming to, your character might lose health and go down a level.

As with all social media, there is a social component whereby your character can interact with others. There's a "Tavern Chat" social forum. You can also join specific groups interested in things such as "getting organized," "finance," "health and fitness," and "advocacy and causes."

You can also set up a habit challenge for others to join—and you can join other people in habit challenges and fight monsters together, which you do by improving your habits. It's like battling virtual monsters by dealing with your metaphoric monsters. It's also a great example of using social media to help us with things we care about and foster personal growth.

So, yes, there's a lot more to social media than the top ten sites. Do some digging, and I think you'll find something off the beaten path that might fit you better than the most popular stuff.

33
Create Your Social Media Checklist

RECENTLY, SOMETHING IN HEALTH SCIENCE was heralded as the "biggest clinical invention in thirty years." Was it a new breakthrough gene therapy? A microscopic robot that can travel throughout the body and zap cancer cells? A microchip placed under the skin to immediately send worrisome changes in blood chemistry to your doctor?

Nope. It was a list of statements with checkboxes on the side.

In 2009, surgeon Atul Gawande published *The Checklist Manifesto: How to Get Things Right*. In this book, he highlights how a simple, humble checklist can change the world. He shows how industries from aerospace to business have used checklists to dramatically reduce the risk of catastrophic errors. Gawande explains how this very simple and inexpensive solution can be applied to hospitals and health care systems to save lives.

These simple checklists can also increase our *personal* ability to achieve our goals in a more consistent and efficient way. Why not put one together for organizing your social media life? You'll be more likely to stick to your goals related to social media and follow through by creating—instead of just falling into—your digital media experiences.

Even just the *development* of a checklist is part of the magic. This is because—as I've mentioned throughout this book—everyone's optimal social media use is going to be different, based on things like personality, preferences, and occupation. Going through the active thought process of setting social media goals will personalize your digital life and balance your technology use.

In this case, the checklist isn't just a set of specific tasks you need to do and then check off, like "take out the garbage" or "call the plumber about the sink." It's about making a list of the ideas this book raises that you want to consider and respond to as you create an optimal social and digital media plan for yourself.

These are big questions you may want to think about before implementing. How many different platforms *do* you want to use? What *are* the ways that you'll avoid negativity? What *is* the right proportion of contacts to designate as "close" friends?

Even if you're not 100 percent sure, review the checklist below and take a stab at creating your own, one that in some way addresses these issues. You can also fill out the checklist directly in this book. It's better to get something down for each topic than to leave it blank. Of course, write in *pencil* (or on a computer), so you can come back and revise. Like all complex behaviors, social media use isn't something that we decide on and then it's done forever. Continued success hinges on our ability to readjust. This list will help you put into practice the social media pyramid by being selective, positive, and creative.

Your Social Media Checklist

Adapt use to my personality

- ☐ Consider conscientiousness:
- ☐ Consider agreeableness:
- ☐ Consider neuroticism:
- ☐ Consider openness:
- ☐ Consider extroversion:

Personalize engagement

- ☐ Amount of daily time I'm aiming for:
- ☐ Daily frequency I'm aiming for:

- ☐ Target number of platforms:
- ☐ Specific platforms that fit me best:
- ☐ Which apps to keep alerts and notifications on:

Decide when to take a break

- ☐ On a daily level:
- ☐ On a weekly level:
- ☐ On a monthly level:
- ☐ On an annual level:

Evaluate my nighttime plan

- ☐ Where is the phone:
- ☐ What time do I stop at night:
- ☐ What time do I start using the phone in the morning:

Establish my contact list

- ☐ My target for a percentage of people I haven't met in person:
- ☐ My target for what percentage of people I'd consider "close" friends:

Minimize negativity

- ☐ How I increase the likelihood of positive experiences:
- ☐ How I decrease the likelihood of negative experiences:
- ☐ My plan for resilience when things go badly:

Reflect on my preferred ways of using social media

- ☐ One-on-one vs. broad posts:
- ☐ Major categories of my posts:
- ☐ Frequency of my posts:
- ☐ Where I fit on the observer-creator spectrum:
- ☐ How I avoid content I'll regret later:

Even if it seems overwhelming at first, ultimately having this checklist should feel freeing, and using it should be empowering.

However, as important as making this list is, I want to address the issue—and even the importance—of cheating now and then. Behavioral scientists know that people who are trying to change a behavior inevitably stray from their goals. Do you just wake up one morning, say you're going to exercise every day, and then do it? Do you decide to eat in a healthier way and then never get tempted by a late-night slice of pizza? Maybe there was one guy, once. But I've never met him.

This is why it's so important to have a plan for *resilience*.

When you learn how to ski, the first thing you do is learn to fall down in the snow without hurting yourself. You practice falling to the left, falling to the right, falling backward, and falling forward. As you practice getting back up from each of these awkward positions, you're strengthening the muscles you need to get back up. You're reinforcing the connections you need in your brain to recover. And you're discovering the emotional challenges involved in feeling helpless and how to calm those emotions when you'll need that ability later.

Therefore, it's important to have a very specific plan for when you *don't* follow your checklist and when you need to revise it.

We can proactively *integrate failure into our tech diet*. That way if you miss a day, or a week, or even a month, you can start again tomorrow. Without that safety valve, a common human tendency is to give up. This is why it's so important to expect lapses—and then to use those lapses to help you recommit to the checklist tomorrow.

Finally, make sure to revise your list from time to time even if things are going well. One of the great things about human beings is that we *change*. Something critical to us today might become less important over time. In the past, people thought that personality characteristics were unchanging, but now we know that's not the case. Let your checklist change with you.

Ultimately, creating a checklist isn't just a way to summarize our own social media plan. It's about *agency*. Creativity itself is about being *in charge* of our own behaviors. As we've discussed, social media platforms conspire to drag us along for the experiences that *they* want us to have. But by committing ourselves to actively creating, maintaining, and adhering to our checklists, we demonstrate our ability to create the digital lives that *we* want to have.

Part 5

Part 5
Part 5
Part 5

How to Live Now

THE BOOK COULD HAVE ENDED THERE.
After all, the main goal was to demonstrate why we need a "food pyramid" for social media and then to explore the three major levels of that pyramid: being selective, being positive, and being creative.

That's what we've done so far. Part 1 explores the cold hard data showing us that social media often doesn't live up to the rosy way it can be portrayed. However, whatever problems social media causes, most people don't want to stop using it entirely. That's often not feasible or desirable in today's world. Instead, we should focus on creating a digital wellness plan to help us foster the positive while buffering the negative.

Part 2 explores the base of the social media pyramid: being selective. This means *actively choosing* the content we want to take in and how we want to use social media. We rule our digital kingdom and can choose what adds to our lives. Thinking consciously about things like how much time we want to spend, how frequently we want to engage, what specific platforms work best for us, and how we're going to schedule our breaks can lead to a healthier and happier experience online.

Part 3 explores the pyramid's middle section: being positive. *Positivity* is especially important when it comes to social media. This means curating our experiences to maximize the good and minimize the inevitable challenges that come up.

Part 4 tackles the top of the pyramid: being creative. Much of this is about tailoring our online experience to our personality. It's also about creatively engaging with social media and making it our own by analyzing and evaluating what we see. Most importantly, creativity in this context is about honing strategies to ensure that we are the ones using social media and not the other way around.

But that only addresses the present—the way things are right now. What about the *future* of social media?

A Changing Social Media Landscape

In 1997, the biggest social media site in the world was called SixDegrees.com. In 1999, it was purchased for $125 million. If, at that time, I told you that it would be shut down the following year, you would not have believed me.

In June 2006, MySpace surpassed Google as the most visited website in the world. At *that* time, if I had told you that MySpace would be gone in just a few years, you would have been similarly incredulous.

In 2005, there was no such thing as Twitter. In 2009, there was no such thing as Pinterest. In 2011, there was no such thing as Tinder. In 2014, there was no such thing as Discord. In 2015, there was no such thing as TikTok. Now there's no such thing as [*Insert next year's big thing here*]. But there will be.

In other words, things change very quickly in the world of social media.

This means that we need to be *flexible* in our approach. We can't focus on one platform, no matter how popular it is right now, because it might not be here tomorrow. We need a system that can apply to *whatever* the future has waiting for us.

That's why the social media pyramid's selective-positive-creative model transcends a particular platform and a particular time in history. Social platforms will boom and die. Thousands of new apps will be created. As marketers continue to vie for our attention, it's a virtual certainty that whatever platform is coming next is going to be just as engaging—if not *more* engaging—than what we have right now.

This is why we need to prioritize principles like being selective. The more advanced technology becomes, the more enticing and potentially addictive social media will become. This means greater risk for interruption of our sleep cycles, even more demand for our attention, and even more difficulty balancing social media with the rest of our lives.

We will also need to focus on positivity more than ever because of how rapidly and dramatically the sheer volume of online data is increasing. It's estimated that the amount of digital data has increased fifty-fold in just the past ten years, and there's no sign of that rate slowing down. The next fifty-fold increase will come with great opportunities for both positive and negative experiences. However, because of negativity bias (see chapter 15), we're all going to have to make special efforts to find, emphasize, and magnify all the positivity we can muster.

I'm reminded of a favorite Lao Tzu quotation that helps me stay on the positive path. It brilliantly demonstrates not only the power of positivity but also the way that very small changes in positive thinking can have a profound influence: "Watch your thoughts; they become words. Watch your words; they become actions. Watch your actions; they become habit. Watch your habits; they become character. Watch your character; it becomes your destiny."

This is why we need to set standards *now* for how we want to live with social media and other technologies. If we begin to pave the path toward digital wellness now, even in small ways, we'll make it easier to continue that path in the future. We'll also inspire digital positivity among those around us, including our friends and family.

One size does not fit all. New platforms will be boons for some people—but nightmares for others. This is why actively creating and personalizing our experience online is a necessary component of our lifelong digital wellness plan.

When new platforms emerge, the specific challenges—and the solutions—will be idiosyncratic to those platforms. But the creative techniques remain the same. These will help us use those new platforms for purposes related to our own dreams and goals, like learning more about the world, sharing our joy with others, expanding meaningful relationships, and lessening others' challenges and grief. Those principles will apply regardless of tomorrow's methods for communicating digitally.

Therefore, even though things are going to change—and change quickly—there's no need to fret. Being selective, positive, and creative is a *flexible* model that holds across time. It's an approach that can help you meet with confidence whatever comes next.

That's why part 5 is important—it focuses on how to apply this approach to many *specific* areas of your life. How does the selective-positive-creative model help us think about using social media in work or school situations? How does it apply to parenting and other family-related issues? How does it apply to different times in our lives and specific times during the year, such as holidays? How can it help us—or challenge us—when we are particularly vulnerable? Let's dig into these and other questions.

34
Work

IT'S BEEN SAID THAT IN many other cultures people work to live, but in the United States, *we live to work.*

Of people employed full-time in the United States in a single job, 31 percent work on an average *weekend day*! And that doesn't mean twenty minutes of answering a few emails. Among those people who work on weekends, they work an average of *over five hours* on each weekend day they work.

People didn't expect it to be this way in the twenty-first century. In the 1930s, John Maynard Keynes wrote an essay called "Economic Possibilities for Our Grandchildren." He predicted that, because of technology and automation, by the year 2000 we might all be working only two days per week. His biggest concern was what we were going to do with all our leisure time.

Obviously, that didn't happen. Not only are many of us working weekends, but the majority of us are *also* working an average of 8.5 hours a day on weekdays. All told, we're spending somewhere around a hundred thousand hours of our lives working, and that figure is increasing as our life expectancies increase and we choose to use the extra time . . . *to work more.*

Work doesn't only capture hours from our lives. In our culture, work often gives us our *identity*. A 2014 Gallup poll asked college graduates whether their job gave them their sense of identity or whether it was just something that they did. About 70 percent of students said that it gave them their sense of identity.

This is not necessarily a bad thing. We can find great joy and fulfillment in work. But when we don't keep a healthy separation between our work and home lives, challenges can ensue.

In the past, work and the rest of life were more clearly separated. On *Mister Rogers' Neighborhood*, Mr. Rogers had a clear ritual he used to start every episode. During the opening song, he walked into his home in a suit jacket, presumably after a day at his office. He then took off the jacket and put on his cozy cardigan sweater to settle into his home life. He also took off his work shoes and put on his sneakers. There was a clear separation of work and home.

Social media, however, has led to our work and personal lives combining in ways we probably wouldn't have expected. Let's start with getting a job. This now involves more than simply filling out an application and having an interview. Now, according to a CareerBuilder survey, about 70 percent of employers review an applicant's social media to screen candidates.

What do we do about this? Do we censor ourselves on social media? Do we change our names frequently?

OCCUPATIONAL VETTING CAN BE YOUR FRIEND

Being selective, positive, and creative naturally helps us prepare for that moment when a future employer vets our social media profiles.

If you are being *selective*—focusing your social media in terms of the time you spend, how frequently you post, and which platforms you post on—you're likely presenting a reasonable picture of yourself that employers are less likely to question. I've been in discussions about potential employees in which the concerns are that "they post too much," "they're posting all the time," and "they're on every platform." This kind of pattern leads employers to question whether that employee will have the time, focus, and wherewithal to get their work done.

Similarly, focusing on people you know well and remaining *positive* and composed in messages will lead to fewer emotional debates that employers might ultimately scratch their heads over.

It's worth noting, of course, that you should still *be yourself.* In some instances, you may be looking for a job that will appreciate your indignant concern over a particular social-justice issue.

Creatively matching your use to your personality may help you appear more responsible and consistent. Also, if you explore alternative platforms, and find a little-known social media site that fits you better than the traditional ones, you may be less likely to appear on an employer's radar in the first place.

A recent survey showed that—among the college admissions officers who used social media to vet their applicants—38 percent said that what they found had a positive impact, while 32 percent said that what they found had a negative impact. So, your social media footprint isn't just a source of embarrassing drunken photos that end up getting you tossed into the circular file. Being selective, positive, and creative can create a social media presence that may be appreciated and admired by people checking up on you.

How you use social media *after* you've gotten a job is also important. People do get fired over their social media posts. If you're being selective and creative about how you post, however, you're probably not going to make some of the mistakes that get people fired—like calling in sick and then posting a photo of themselves bungee-jumping.

Also, remember, you aren't the only one posting about your life. Imagine you are supposed to be in class or working, but you decide to join a group playing hooky at a pancake house. It only takes a moment for a friend to snap and post a photo of you happily enjoying yourselves. It used to be the telltale blueberry syrup on your lip that gave it away, but today the photographic evidence can be more permanent.

The language people use, whether in the office or online, can also get them fired. In many organizations, management systematically searches social media for any information about the organization and its employees, sometimes using complex machine-learning algorithms. While we are all entitled to our opinions and to free speech, companies are also allowed to fire employees for violating company policies. Exaggerated statements made in jest—such as saying we want to "kill" our supervisor or "destroy" the company—may not be taken as jokes. Out of context, a sarcastic social media post can lead to a pink slip or worse, such as charges of making terroristic threats.

Social media makes it increasingly hard to compartmentalize our work and home lives. By "friending" a coworker, that person can access and share everything you post, but even without that, depending on your settings, your social media information might be available to anyone who cares to find it.

Creating *positive* content is also useful in this case. No job is perfect, just like no person is perfect and no institution is perfect. But if we use social media to perpetuate the positive, we are less likely to make any statement that might be misread or cause problems.

The multiple discussions in this book around *creativity* will also give you a smoother path when it comes to keeping things straight at work. For example, communicating with people one-on-one or in small groups can keep your interactions more private and less discoverable. If you're in the habit of broadcasting to the whole world only when it's absolutely necessary, you'll be less likely to post something that can come back to bite you.

It turns out that work-related platforms may also carry some of the same challenges that social platforms do. One of my students, Jacquelynn Jones, and I published a paper a few years ago focusing on LinkedIn, and we found that frequent exposure to it led to some of the same social-comparison concerns we had seen with other platforms.

We also found that these risks were different depending on your demographic characteristics. For example, if you were unemployed, the more time you spent on LinkedIn was related to significantly more depression and anxiety. People in lower socioeconomic brackets were also highly sensitive to spending more time on LinkedIn.

This makes sense. If you're wealthy and well-employed, lots of time on LinkedIn may help you feel better about your station in life and your overall success at work. But if you are unemployed, poor, and searching, a lot of time on LinkedIn might make you feel like you are not measuring up.

The irony of the situation is that, if you are unemployed and not well-off, you very well may *need* to spend time on work-related social media platforms forming the right connections to help you move to the next level. However, by being *aware* of the emotional dynamics, they may not affect you as much.

Then, before you know it, you may be sharing the good news of your new job on social media.

35
Family Life

TEN-YEAR-OLDS IN THE UNITED STATES can name more brands of beer than they can past US presidents.

Although the study that supports this fact was conducted in 1988, I suspect that its findings are still true today. Every Sunday, while watching football games with their families, millions of kids across the country see compelling portrayals linking joy, friendship, social status, and humor to Bud Light, Coors, and Heineken. Young people soak up information from billboards and point-of-sale ads about Michelob Ultra, Busch, and Stella Artois. And aggressive social media campaigns promote Budweiser, Dos Equis, and even craft beer companies like New Belgium.

But for some reason, there just isn't the same marketing power attached to Martin Van Buren, Zachary Taylor, Rutherford B. Hayes, and Gerald Ford.

Starting in 2017, there was an extensive Bud Light mass-marketing effort set in the *Game of Thrones* universe where commercial characters toast—using Bud Light, of course—with the nonsensical words "dilly dilly." The silliness of the words and the tie-in to an extremely popular television program led to teens debating ad nauseum the meaning of "dilly dilly" and purchasing a variety of T-shirts emblazoned with the phrase.

On the other hand, those teens are far less likely to wear a T-shirt featuring Grover Cleveland—or even know that he was an American president, and in fact, the only president to have served two non-consecutive terms. It's not that I blame these kids; they're simply inundated with media portrayals of beer and comparatively less exposed to history education.

This exemplifies, however, a major reason why parents are concerned about the influence of social media on their children. Kids and teens are growing up in an age when technology, apps, and social media—many of which didn't exist even a couple of decades ago—form the basis of most of their waking thoughts. If a kid can learn the names of beer brands from a few commercials, what else are they subconsciously learning in today's digital world?

On social media, advertising isn't random. Persuasive content is *targeted* to us based on our likes, our personality characteristics, and our past behavior. The ads that appear are carefully curated to appeal to our specific needs and desires, no matter what age we are. They are designed to seep into our thoughts, and youth are especially vulnerable to these psychological processes. It's no wonder that most parents are concerned about privacy and manipulation of their children in the era of social media.

Parents are also concerned about vulnerable youth being lured into unwanted communications. The FBI estimates that a half-million sexual predators are online every day, and more than half of victims are between twelve and fifteen. This is a particularly sensitive age psychologically and emotionally. It's also when youth are just figuring out how to use social media, which can lead to errors in judgment. Cyberbullying is a related concern of parents—it's estimated that about 36 percent of youth feel they've been the victim of cyberbullying, and for about half of those individuals, bullying has occurred in the past thirty days.

The *permanence* of information on social media is another worry. Kids say embarrassing things all the time, and it's usually no big deal. It's heard by few people and soon forgotten. But an embarrassing or offensive tweet can live forever and be seen by everyone. The ability to take screenshots means that even when posts are taken down, they can be recorded and reshared to provide fresh embarrassment at any time.

Finally, parents are deeply worried about the sheer volume of time their kids spend on social media. It used to be that, to control the amount of media kids watched, you just turned off the TV during those few hours of prime-time, but restricting access to 24/7 smartphones is trickier. Phones are used for more than social media, and simply saying, "Don't use too much social media," isn't an ideal strategy. In many ways, it's not a fair fight because parents are up against an army of marketers and persuaders supported by billions of dollars. Despite this challenge, I think that the social media pyramid can empower us in our efforts.

EMPOWERMENT: A PARENT'S SECRET WEAPON

The social media pyramid emphasizes positive phrasing for a reason. It doesn't warn us about what not to do. It names what we *should* do: Be selective, be positive, be creative. This empowers us to actively select what is best for us, and empowerment is a parent's secret weapon with their children. This gives kids the agency they desperately seek.

It is a normal part of development for young people—especially those ages eleven to nineteen—to exhibit "reactance," which is the urge toward the *opposite* of what someone tells us to do. A major purpose of adolescence is to separate from parents and other adults, so reacting against authority is expected.

That said, kids differ in their level of reactance. Not every child is automatically contrary, and even kids who generally resist parental advice will sometimes cooperate. It also doesn't mean that parents should not take a stand; while a young person is rolling their eyes at something that a parent says, they're often also filing that away as something that they ultimately will listen to and believe in.

Still, we need to be thoughtful about how we approach youth. A good example comes from the DARE campaign, which began in the early 1980s. DARE stands for "Drug Abuse Resistance Education." This well-intentioned educational campaign involved police officers directly interacting with youth to tell them about the perils of using alcohol, taking drugs, and resorting to violence.

This effort had mixed results for several reasons. One was reactance; many kids resisted being told what not to do by police officers, the ultimate symbols of authority. DARE also wasn't optimized for its key target audience: the sensation-seeking, rebellious kids who were more likely to need this kind of education in the first place.

Traits like sensation seeking and rebelliousness are often inborn. Some people have brain chemistry that makes them enjoy taking risks other people find terrifying. This is why roller coasters thrill some people but make others just feel sick.

This led to an interesting—and problematic—paradox in the case of DARE. It worked just fine for kids who were *not* risk-takers by nature. But these were not the kids who were most likely to experiment with drugs in the first place. The teens who were at highest risk were given a blueprint for the riskiest, most rebellious behavior they could engage in.

Indeed, later studies about DARE's effectiveness showed very mixed results, and some studies found that, compared with people who were not exposed to DARE, those who were had actually used *more* drugs.

DARE provides a valuable cautionary tale for parents speaking to their kids about the dangers of social media use. Parents want to help kids recognize the dangers and challenges that can come along with too much social media use, but they don't want to spark reactance against themselves.

However, by providing positive messages helping kids to improve their own experience—by being selective, positive, and creative—parents can steer their kids' reactance and rebelliousness toward the entities that are *truly* manipulating them: social media companies. This means minimizing "rules" and speaking more about *choosing wisely* how to use social media—including how much time to spend online, how frequently to log on, which platforms to use, whether to use social media before bed, selecting the people they interact with, and so on. Within parameters, and in ways appropriate for their age, allow children to choose in ways that empower them.

Further, discuss with kids how they can adopt a *positive* approach to their own use, and to be alert to the negativity social media can breed. There's a stereotype that adolescents are negative toward everything. However, many psychologists think that this is less about being truly negative and more about being *bored* and *confused*. Think about it. You're a developing kid. You've been quite happy for the past decade watching cartoon movies, having birthday parties with frilly cakes, and collecting Pokémon cards. Then, all of a sudden, none of that interests you anymore. What do you do now? What *replaces* all of that? Social media provides adolescents an opportunity to "try on" new identities and ways of being, to explore who they are and who they want to be in new contexts. Giving youth a mindset of *positivity* as they try on new identities can be life-changing.

Here's an example. About a decade ago, my research team and I made a documentary film with kids called *AD IT UP* based on the media literacy model described in chapter 27. We brought teens into a studio and taught them the principles of media literacy as they relate to tobacco use. But instead of focusing on the negative—like how smoking is harmful to them—we leveraged their natural teen energy, inventiveness, and enthusiasm in a *positive* direction. Their task was to make video messages aimed at teaching other kids how to think critically about advertising and marketing related to tobacco. The students made remarkable products and left that experience with a true sense of optimism and empowerment, and they also left with lower intention to use tobacco in the future.

Ultimately, encouraging kids to be *creative* on social media might be the easiest and most effective strategy. In addition to separating from parents, another natural purpose of adolescence is to *develop uniqueness*. So, the idea of being *creative* and *different* in how you use social media might resound particularly well with youth. For example, they may have a good time looking for unique platforms. They may be easily engaged in thinking about their own personality and matching how they use social media to enhance their uniqueness. Many kids like the idea of being active creators and analyzers of content instead of passive observers. They like taking charge of their own experience and tailoring it to what suits them best.

Parenting has been described as the hardest job. Now the advent of social media has introduced additional—shall we say—complexities. Hopefully, the ideas we've discussed around being selective, positive, and creative can be secret weapons for parenting in the age of social media.

36

Tech Is for the Young, Right?

THE FASTEST GROWING GROUP OF social media users in the United States are those ages sixty-five and up.

However, there's been very little research around social media use in this group. On the one hand, as with young adults, we might expect older people exposed to more social media to feel more depressed—because of things like social comparison or taking too much time away from more valuable in-person experiences.

On the other hand, there could be a completely different relationship between social media and emotional health in this age group.

In modern-day Western society, many older individuals are physically isolated. In developing nations and in the East, older people tend to live with extended family members, but in the United States, widows and widowers tend to live alone. A recent Pew research study found that 50 percent of people who live in Asia over the age of sixty live with extended family, while this is only the case for 6 percent of people ages sixty and over in the United States.

Therefore, social media might function very differently for elders than it does for young people. Among older individuals, it may offer more unique opportunities to form connections that might otherwise be missing. As we get older, retire, and settle down, we may have fewer opportunities to meet new people and connect with old friends. Given these cultural realities, social media might provide a valuable gateway for keeping in touch and having a social life in old age.

THE POWER OF A PURPLE STUDEBAKER

Online and digital content like music and videos also may lead to powerful positive experiences for many older people—not only to relive memories of the past but also to connect with others.

My kids and I have been volunteering for the elderly for about fifteen years now. My son started coming with me to assisted-living centers when he was a toddler. Obviously, he was the star of the show, not me. When we played on the guitar and sang "Home on the Range," "Summertime," and "Keep on the Sunny Side," there were magical moments. The staff told me that people who had been nonverbal for months suddenly sang along. Now my kids can do these performances themselves, and I know it's one of their most gratifying experiences.

Social and digital media use among the elderly similarly can deliver lost gems of the past, whether through music, movies, or stories.

In my role as dean of education and health, one of the areas I oversee is the Osher Lifelong Learning program, which focuses on bringing nondegree coursework to older individuals in the community. The 2020 coronavirus crisis hit this program hard. Many of our hundreds of members and attendees relied on Osher for their social and enrichment experiences. Most members hadn't ever used platforms like Zoom, so pivoting to remote learning wasn't going to be an easy sell.

But we did our best. Instead of my usual address to the community, I helped everyone get online and created an interactive digital trivia game focusing on the 1950s to the 1970s. Eventually, with the help of digital media—photos, videos, and unique interactive elements of the delivery platform—we relived moments in entertainment, politics, and history that made people remember and smile. They saw Audrey Hepburn in her prime, saddle shoes, the Everly Brothers, a purple Studebaker, and much more.

As a society, we need to think much more about how small experiences like this—and social and digital media in general—might help connect older individuals and improve their lives. In addition to isolation and marginalization, this population faces important concerns about health care and economic well-being in the future.

The number of individuals sixty-five and older is projected to *double* in the next few decades. By 2035, people in this age group will, for the first time, number more than those under eighteen.

This demographic shift may lead to problems. For example, will the health care system be able to keep up? It's projected that, by 2030, the percent of *nonworkers* (mostly retired individuals) who depend on workers will be 70 percent. Soon after this, in 2035, both Social Security and Medicare are projected to become bankrupt. Does this mean that fifteen years from now a huge swath of the country will have no health care and insufficient retirement—aside from the ultra-wealthy?

Social and digital media, of course, can't reverse these major demographic, societal, and financial trends. But social and digital media may help in a few ways. First, social media can provide connections that may have been impossible otherwise. My wife's mom has been living with us for about the past two decades. Using digital communications, she is now more connected than perhaps any of the rest of us. She has no fewer than four regular weekly knitting dates with people from all over the United States. Each event may not be as powerful as an in-person experience might be, but they certainly help fill her time and mind with stimulation and warmth. She's also able to use these platforms to visit with distant and scattered family members—for key life events or just to catch up.

Social and digital media might also improve some larger societal issues. In terms of health care, the growth of telemedicine may help older populations significantly. The ability to come into the clinic physically for a simple medication check has been a substantial barrier in the past. But visits over visual platforms can be convenient, efficient, and cost-effective.

Unfortunately, social media increases the risks for seniors to be targeted by various scams. The elderly are concerned about health care, so they're targeted by health insurance scams and counterfeit prescription drugs. They're concerned about their retirement funds—and being able to help their children and grandchildren—so they're targets of fraudulent investment schemes, lottery and sweepstakes scams, mortgage scams, and more.

This is why we need to protect and empower our elders in these areas.

At the University of Arkansas, we're facing up to scammers by working with a community-based organization called SALT—"Seniors and Law Enforcement Together"—which presents free senior safety programs.

Using the social media pyramid, and being selective, positive, and creative, will also help seniors navigate the opportunities and liabilities of social media.

When we help older folks be selective about platforms, it reduces the risk that they'll be targeted. By using fewer platforms, they are more likely to master those select systems, making them less likely to be manipulated. We can also help seniors to be selective *within* the platforms they select, sticking to features and methods of interaction they feel most comfortable with.

Learning how to use social media positively may be even more important among older people than it is for young adults. Older people might have naïve expectations about social media, thinking, *What a big group of wonderful friends!* Then as they experience flaming, trolling, and other negative behavior online, they may need help coping with these challenging issues—in part by training them to keep interactions positive to avoid unwanted attention or regret.

Finally, helping seniors develop media literacy—the ability to critically analyze and evaluate what they see on social media—will be key to recognizing and dealing with scams, turning off unnecessary alerts, and using social media in a more empowered way.

In short, growing old in the United States is hard. Especially as society increases its reliance on and infatuation with new technologies, older individuals who did not grow up with these technologies can feel marginalized and disenfranchised. At least in part, the best solution may mean embracing the old maxim "If you can't beat 'em, join 'em." With appropriate support, older individuals can learn to leverage the tools of digital technology to improve their lives through connections with social services, friends, memories, and family.

37 The Holidays and Consumerism

HUNDREDS OF THINGS MIGHT HAPPEN to us over the course of a day, many of which might make us feel disgruntled, ashamed, irritated, or even depressed. Yet we post that one thought or picture that makes our overall life seem rosy. Along with everyone else, we constantly curate our brand, image, and content to showcase the best possible version of our lives.

This curation is not unique to social media, and it's not a bad thing. According to the National Science Foundation, we have between twelve thousand and sixty thousand thoughts per day, and 80 percent of them are negative. If we didn't curate and filter these thoughts, we'd all be wallowing in a sea of negativity.

However, even if curation isn't itself a problem, the self-comparison that arises out of it often is. As we've discussed, we need to constantly remind ourselves that social media presents a skewed version of reality.

One time when this skewed reality often becomes the most acute is during the holidays. While this issue applies to any holiday, the winter holiday season is often the most challenging.

In my eighteen years in clinical practice as a family doctor, I learned about the massive disconnect between the external unadulterated joy my patients thought they were *supposed* to feel during the holidays in comparison with the challenges they *actually* went through at that time.

For various reasons, many people regard the holidays as the worst time of the year. One theme is grief. At holiday gatherings many people notice and mourn the absence of people who have died, whether recently or in the past, and whether that death was tragic or expected.

Another theme is money. People worry they can't afford to get presents for everyone, and if they don't get someone a present—or a present isn't expensive enough—what will everyone think of them?

Then there's travel. People often don't want to drive eleven hours or fly across the country only to have their plans foiled by winter weather and lost luggage.

Social media can magnify all these issues. For one thing, the usual curation and filtering of social media tends to increase substantially during the holidays. Even while people's internal lives are racked by stress and anxiety, their feeds become that much more glowing. Everything is supplanted by holiday photos, cards, and events. This increases our vulnerability to feeling like we don't measure up.

It's also not a small thing that *commerce* ramps up massively during the holidays. The National Retail Federation estimates that holiday sales represent about 20 percent of total annual sales, and it can be even more for some retailers. This shopping binge between Thanksgiving and Christmas leads to a flurry of stressful situations on social media.

In some ways, e-commerce can reduce some of the stress of shopping in physical stores, since it means no parking hassles and no fighting with crowds. However, e-commerce also can make it easy to overbuy, spend too much, and have buyer's remorse. While being able to shop 24/7 is convenient, it means we can always be shopping even in the privacy of our bedrooms.

The *targeting* afforded by e-commerce can also increase stress. In the 2002 movie *Minority Report*, based on a 1956 short story by science fiction master Philip K. Dick, the billboards *have* eyes and *examine* eyes. Billboards use retinal scanners to identify passersby, and then they personalize subsequent marketing messages. At one unsettling moment in the story, the protagonist tries to remain as anonymous as possible to flee the police who are after him. Yet a billboard scans his retina, cheerily calls him by name, and suggests that he purchase another shirt like the one he recently bought. Does that sound familiar? Maybe *billboards* don't do this—yet—but the online ads you see while browsing social media—and the coupons you receive in your email—are based on multiple data points, including your prior purchase history on those sites. Netflix prides itself on recommending entertainment based on your viewing history, preferences, demographics, and prior ratings. With the information they have, marketers have turned science fiction into science fact.

THE PREGNANCY PREDICTION MODEL

Targeted marketing has come a long way, even over the past couple of decades.

Consider the story of the Minneapolis man who angrily strode into Target to ask why his teenage daughter was receiving Target coupons for baby clothes, cribs, and other items for infants. The Target manager apologized profusely and then called a few days later to apologize again. But by then, the father was contrite. Since his visit to the store, he had learned that his high school–age daughter was indeed pregnant.

How did Target know that his daughter was pregnant? It was simple, actually: data mining and algorithms. Every person has a "Guest ID." When anyone signs up for a credit card, or just uses a credit card, data like their age, sex, and zip code are attached to the Guest ID. Then, when they buy something, the product, time, and location of that purchase get added to their Guest ID. Taken together, these data tell companies like Target a lot about a person's life. In this case, Target scientists had created a "pregnancy prediction score" based on customers having bought twenty-five different things—like prenatal vitamins, a large purse that can double as a diaper bag, and a kind of cocoa butter that purports to reduce pregnancy stretch marks.

Once a woman reaches a certain threshold on the pregnancy prediction score, she starts getting pregnancy-related coupons. These coupons arrive via email, mail sent to your house, and even printed at the end of *other* store receipts.

It can feel disconcerting to recognize how much of a permanent record exists from our purchasing, browsing, and clicking patterns. This profiling is also unsettling because it doesn't know when to stop. Twenty years ago, you could window-shop without worrying about being followed home by the department store. Now, however, we've all had the experience of innocently typing on a whim "Best hybrid SUVs" into a search engine, and then getting inundated—for weeks, if not months—with free trials, suggestions, campaigns, slogans, and compelling images. Maybe this won't prompt us to buy a car if we don't really want one, but being persuasive is what marketing is all about. We often buy what we don't need, or spend more than we want, because an army of designers, psychologists, and marketers are doing their jobs well. And this marketing pressure becomes even more dramatic around the stress of the holidays.

So be aware, especially at holiday time, about how much everyone is trying to influence our experience of the season. Then be selective, positive, and creative with your digital media consumption—and this just may help your holidays become more of what the holiday season is actually supposed to be about.

38
When You're Vulnerable

IT'S AN INTERESTING PUZZLE. The times we are vulnerable are the times we tend to reach out via social media. But those are also the times that social media may present the biggest challenges for us.

Several years ago, I read a fascinating book called *Why We Get Sick*. It discussed that there may be good reasons—in the evolutionary sense—for why we get certain illnesses.

When I was working as a family-practice doctor, many of my pregnant patients had severe nausea and vomiting daily. What a terrible irony, I thought, that these poor women were trying to stay nourished for two, but they couldn't even keep enough down for one.

The authors of *Why We Get Sick* explain that morning sickness makes sense from an evolutionary perspective. In prehistoric times, women with the urge to expel everything they ate during pregnancy probably had a survival advantage, because a lot of the things they were eating might have been bad for their babies.

Sure enough, today, pregnant women are given an extensive list of things they aren't supposed to eat: undercooked meat, raw eggs, alfalfa sprouts, soft cheeses, alcohol, caffeine, most medicines, undercooked fish, and so on. Nausea during pregnancy became a built-in mechanism to make sure women avoided these and other problematic foods.

What about *depression*? Could there possibly be any "good" reason for people to feel despair?

Potentially, yes, from an evolutionary perspective. Imagine a hunter-gatherer tribe in prehistoric times. After running out of food in one area of the forest, the tribe moves to a different part of the forest. Everything seems great, until one day, someone in the tribe goes out to hunt and never comes back.

What do you do? Is it better to be fearful and experience dread, or to be unafraid and unaffected? A happy tribe might continue hunting in the area only to encounter the same family of aggressive bears that killed the previous hunter. A fearful tribe might stay put while mourning their friend, play it safe, and avoid the same danger.

In other words, even depression has a basis in our evolutionary biology. Depression causes uncomfortable feelings, but that doesn't mean it serves no purpose. It may get triggered inappropriately—when there is no danger—but it's in our genes for a reason.

SOCIAL MEDIA: THE NEW KID ON THE EVOLUTIONARY BLOCK

But here's the problem with social media: It hasn't been around for millions of years. We haven't had time to adapt to it or to develop natural defensive instincts that protect us from its influence when what we click and how we consume are damaging to us.

We can recognize self-protective instincts in our everyday lives. It isn't only helpful when we're faced with predators. Illness, for example, makes us want to curl up in bed. This is important, since our bodies require huge amounts of energy to make antibodies and fight viruses and bacteria. Slowing down saves our energy so we can heal. Would it be better if we had an instinct to go out and run a marathon at the time of our illness? Probably not.

We have the same reaction if we get into financial trouble. Say we start a business such as a restaurant, and within a year, it starts to fail. Is it now the best time to smile it away and go for a *bigger* loan at a *higher* interest rate? Probably not. It's probably better to regroup, limit the damage, and try to slowly and responsibly get back in the black.

The same is true in romantic relationships. If a long-term relationship ends, and two people part ways, they typically take time to grieve their loss, evaluate what went wrong, and heal emotionally. They generally don't shrug and start dating other people the very next day, assuming the next person they meet will surely be "the one."

So why don't these same self-protective instincts kick in to deal with the challenges of social media?

One reason, of course, is that if we're in pain, social media can be a natural place to turn to for community, connection, and emotional support. I experienced this during the aftermath of the Tree of Life massacre in 2018. My friends and I craved information and community. We became glued to our social media feeds. That connection was helpful and allowed us to support one another even when we couldn't physically be together. We were able to celebrate and share hope—like when we learned that one of our beloved members was indeed still alive, or when the Muslim community stepped in and paid for all eleven funerals of their fallen Jewish neighbors.

But as this book has discussed throughout, social media platforms are complicated forums for connection. We don't always respond to posts and interactions in positive ways, especially when we feel vulnerable. For instance, if we post about being sick in bed, and then spend most of our time scrolling through social media to see who responds, we may end up feeling worse. While some people will respond with sympathy, they may not be as sympathetic as we'd like, and we might wonder why others didn't respond at all.

Social media platforms also monitor and respond to how we use them. As they try to monetize our clicks and communications, we can be inundated with advertisements and pushed to explore "related posts." These systems are often designed to leverage our vulnerability, which means that at precisely the moment we most need trust and support from others, we are often being manipulated, usually to buy something that will solve our problem. Did you post with a word like "sick," "cold," "cough," or "fever"? Don't be surprised to be fed advertisements about miracle cures that will rev up your immune system into such a miraculous state that you'll never be sick again. Maybe that miracle cure will indeed fix our cold, or maybe it will just create a $14.99 monthly charge that will take months to get off our credit card.

What about financial misfortune? It's easy to log on and engage. You're already glumly examining your accounts and the red ink all over your spreadsheets. But is that really the time for you to be seeing one of your acquaintances from high school showing off their pictures from a trip to the South of France, something that you can no longer afford? And is it the right time for you to be seeing lots of ads for high-interest loans and get-rich-quick schemes?

Finally, because most people curate their social media to reflect their best, turning to social media for support when we're in pain can lead to particularly acute negative social comparisons. If we just ended a romantic relationship, how will we feel scrolling through photos that celebrate everyone else's loving partners and spouses? We might only feel more ashamed and lonelier.

What about our platform's "relationship status" feature? We may instinctively want to change our status from being "in a relationship," but when and how? Should we wait until we feel less vulnerable or do it right away—and do we pick "single" or "it's complicated"? Do we really want to announce our breakup to the whole world and risk unwanted attention from people we don't know well who are asking questions we don't feel ready to answer?

Social media *can* be genuinely helpful when we are going through a hard time, but we shouldn't just log on to a platform, share our pain without careful planning, and mindlessly absorb whatever is in our feed.

During vulnerable times, it is even more important to stick to our tech diet of being selective, positive, and creative.

So, during trying times, it's especially important to communicate one-on-one and surround yourself with positive content. Focus only on platforms that empower you, and consider how to individualize your social media use according to your personality. Approach what you see critically.

Ironically, being in a vulnerable state can either influence us to be more careful and self-protective—or it can lead us to throw all our rules out the window. Lean in to our adaptive, self-protective instinct in difficult times. Take a cue from our hunter-gatherer ancestors. When losing a friend to a bear, it's not a bad idea to curl up and regroup before reemerging in a more careful, thoughtful way—especially when it comes to using social media.

39
Mindful Social Media

IN ORDER TO TRANSFORM YOUR experience on social media, the last thing I suggest is to practice "mindful social media." What is that, and how do you do it?

At face value, this concept may sound like an oxymoron. Social media can seem like the bitter enemy of mindfulness. Every corner of every display is packed with stimulation designed to vie for our attention. Videos, pop-ups, pings, and alerts constantly distract us. Advertisements litter the social media interface. How can we practice mindfulness with all this going on?

These advertisements are designed to catch our interest—each one of us individually—based on our past searches, purchases, and activity. Scrolling through social media, our minds usually whir with thoughts, reactions, considerations, emotions, reminders, and assorted bric-a-brac. We consider what to buy and what not to buy, a friend's heartfelt fundraising plea, the compelling new virtual event series someone is attending, the instant chat requests waiting for us, all the birthdays happening that day, the video suggestions—all of which may be encountered before we even get to our main feed.

We know all this. So instead, let's start by considering mindfulness. One simple definition of mindfulness is "a state of focused attention."

Note that this definition doesn't say anything about sitting in the lotus position, being on the top of a mountain, listening to gongs, or wearing yoga pants. Anyone can practice mindfulness anywhere, but it does take practice.

One common technique for introducing beginners to mindfulness is to take a small object, like a raisin, and simply focus on it for a period of time with all your attention. For five or ten minutes, try to focus on the raisin and think of nothing else. You can start by looking at the raisin—exploring its crevices, noticing the light reflecting off certain parts of it, and following the little sections of it that bulge out.

Eventually, you explore the raisin with all your senses. Feel it at the tips of your fingers. Notice if it feels cool, warm, or in between. Notice its texture: Is it rough, smooth, squishy, or something else? You can even ultimately taste the raisin, the usual way we experience raisins, but only after comprehensively taking in all aspects of its raisinness.

While you do this, of course, other thoughts will inevitably appear in your head that are not exactly about the raisin. You may think, "Gosh, I really dislike oatmeal raisin cookies. Why on Earth would people make those instead of good old chocolate chip cookies?"

When you notice one of these *intrusive thoughts*, you simply recognize it, acknowledge it. *Wow. That's interesting. I was nicely focused on the raisin and all of a sudden I started thinking how I feel about various kinds of cookies*. This is where the magic comes in.

You gently *nudge that thought aside* and go back to where you were, observing the raisin. Inevitably another thought will intrude—maybe listening to a dog barking in the background—and you push *it* aside. The next intrusive thought is about something you have to do for work. But you don't judge yourself harshly for not being able to focus—just keep nudging aside and coming back to the raisin.

This is the practice of mindfulness meditation: Focusing your attention, noting when intrusive thoughts arise, and then returning to your original point of focus.

The first time you do this, thoughts might interrupt your focus every couple of seconds. But just like anything else, if you practice and do this exercise regularly, you will hone your ability to refocus your attention whenever it gets interrupted, and your skill with mindfulness will improve. This is no different than learning how to play an instrument like the clarinet. The first time you play, you might sound like a panicked goose. But even just a week or two later the goose might not be *quite* so panicked. A couple of weeks after that, your rendition of "Twinkle Twinkle Little Star" is only *slightly* painful to your friends and family, and eventually you're actually making music.

Some say that mindfulness meditation is even more difficult to master than something like the clarinet. This is because, no matter how long you practice and how good you get, you will *always* have intrusive thoughts. This is why people call mindfulness meditation a "practice"—not a skill you master and are then done with.

One reason to develop mindfulness is because it's good for us. *Really* good. Studies show that mindfulness can improve sleep, reduce anxiety, and help us achieve goals like improving work output, losing weight, or writing a book.

But another reason to cultivate this humble skill—that of being able to come back to your awareness of the present moment—is that you can use it to dramatically improve your experience during the hours you spend using social media.

In some ways, social media is the ultimate test. If we can practice mindfulness while using social media—to avoid mindlessly scrolling and all the negative consequences that can lead to—we can learn to be mindful in just about any situation.

HOW TO BE MINDFUL ON SOCIAL MEDIA

Most people use social media an average of two to four hours per day, and even more than that on certain days. This is why, however challenging it might be, it can be a good habit to use our time online as an *opportunity* to practice mindfulness.

There is a value to honing this ability in an *unripe* context for mindfulness. Trying to remain mindful while on social media can improve our skills dramatically, even more than if we're in an ideal situation. It's sort of like the way baseball players, while they are on deck, practice by swinging bats with weights on them. That way, when they step up to the plate, working with a regular old bat feels like a breeze. If you can somehow keep your attention focused while navigating the frantically stimulating world of social media, it might make it even easier to use what you've learned in the actual world.

But how can we possibly focus with so much noise online?

Practicing mindful social media requires exploring your experiences in more complete ways, such as zooming in on one focal point to experience it more fully.

By default, social media is very much about focusing on *everyone else*. Your attention is being rapidly whisked from post to image to icon to emoji and back. Your attention is everywhere and nowhere all at once. That's sort of like experiencing raisins by devouring them by the handful.

Instead, mindfully zoom in on each post and focus on it exclusively, noting its intricacies in the same way, metaphorically speaking, you would a single raisin. Focus on the person's message, take your time, and explore it completely. Notice the way it makes you feel. Notice when your attention is pulled away and treat it like an intrusive thought while meditating. Gently bring back your attention to whatever you were focusing on.

There are two major benefits to processing social media in this way. One is that we're more likely to truly *experience* what someone has shared. Instead of just giving it a cursory glance and moving on, we're more likely to develop compassion and empathy for that person. This is then more likely to lead to a deeper relationship. Something that could have been superficial has become *meaningful*.

The second benefit to experiencing our platform in a more mindful way is that we're less likely to fall into the traps of social comparison and negative reactions. If we notice feelings like envy or depression, or any instinctive and subconscious reaction, we can treat it like an intrusive thought: See it, acknowledge it, then put it aside. We can say to ourselves: *There I go again, feeling envious because my friend has posted about their nice vacation. I wish I had a nice vacation to post about. But I can still be happy for my friend.* When we make our feelings and anxieties conscious, we are less likely to be controlled by them.

So, practice mindfulness, both offline and online, and don't worry if you don't have yoga pants or a gong. Focus your attention, and when it drifts, come back to what you were doing. Let your focused attention help develop compassion for others—and for yourself—even within our frantic digital world.

Conclusion

Just as our bodies grow flabby if they are not challenged by hard work and exercise, our souls can grow flabby if they never encounter adversity.

—Harold Kushner

WHEN I SET OUT TO write this book, I didn't expect it to be this long.

I was energized to share ideas I had been studying and contemplating for two decades. I thought it would be rather simple. I thought that I'd put together a few key pearls for optimizing social media use in today's world, and that would be that.

But the process of actually writing this book changed the material, and it changed me. For one thing, I gained greater perspective on the *complexity* of these issues. We can't just think about *how much* we use social media. We also need to think about *who* we're interacting with. *Why* we're using social media in the first place. What particular *platforms* we're using. *When* we're using them. How our unique set of *personality characteristics* and *experiences* meshes with all these things. And more.

Despite this complexity, I also know from being a student of behavioral science that we need a relatively *simple and flexible* way of packaging all these messages. As humans, we just don't have the bandwidth to take in everything at once.

One of my favorite illustrations of this is a classic paper written in 1956 by George Miller, a Harvard psychologist. The title of the paper is "The Magical Number Seven, Plus or Minus Two: Some Limits on Our Capacity for Processing Information." It's one of the most referenced papers of all time. In brief, Miller describes how he concluded—from lab studies and theory—that the average human can only hold about *seven things in short-term memory.*

What? If this is true, how do we remember the hundreds or thousands of things that we *seem* to hold in short-term memory? Simple: through organization and branching. In our mind, one thing is a branch off another thing, which is a subcategory of yet another branch. So, the power of exponents helps us access thousands of memories.

An implication of this work is that we need things to be *simple and organized* for them to be useful to us. While no system is perfect, being selective, positive, and creative is a good start when it comes to optimizing our digital lives.

Selectivity is something that we've had to learn relatively recently. Only a couple of generations ago, there were just three television programs available to watch at any given moment. Today, there are over five hundred *active* scripted television programs, and that doesn't count all the *archived* material you can watch immediately, at any moment of the day.

Once upon a time, books were only physical objects you owned or borrowed from a library. Now, about fifty million books are available on Kindle alone, not to mention all the newspapers, magazines, articles, and formerly printed material you can access electronically. In the 1980s, data was measured in kilobytes. Now it's gigabytes and terabytes, which are *one million* and *one billion* times as big as kilobytes, respectively. It won't be too long before we're measuring in petabytes, exabytes, zettabytes, and yottabytes. I'm not making these terms up. Each represents a thousand times as much as the previous term.

Today, there's no shortage of information. The biggest shortage is our *attention*. But we're not great at parceling out that precious commodity. How many times do we binge a TV series only to wonder what else we could have been doing with that time? In clinical practice, I had to let a lot of people know that they only had a certain amount of time left to live. Nearly all of them expressed regrets about how they had been focusing their attention. Almost no one said they were going to keep living exactly the way they had been before.

That doesn't mean we are to blame. On social media, and in general, we're up against a lot. The marketers and corporations that vie for our attention are smart, well-resourced, and effective. This is why enhancing our selectivity is so important. And that's true whether we are being selective about our social media habits or about other aspects of our lives.

Positivity is another skill that applies equally to our social media lives and our world in general. As this book discusses, positivity is not about being a Pollyanna and ignoring difficult things that happen. Instead, it's about maintaining *perspective* in the face of difficulties. Today's digital technology has made this harder than ever. Whatever terrible things are happening in the world at any time, news of those events—along with video and commentary—is immediately available in our bedrooms. Blissful ignorance of world events is no longer possible, even if those events have no immediate impact on our day-to-day lives.

For this reason, we need to find a balance. We need to be positive to work toward achieving a greater and wider perspective. If we don't, negativity often will win.

Creativity—the last aspect of the social media pyramid—is also something that we need when using social media *and* to improve our lives in general. In essence, if we're creative in terms of how we analyze, evaluate, and critically think about our world, we'll make discoveries about how we spend our time and energy online, and this will ultimately improve our lives.

Also, remember that the strategies discussed in this book might create some pushback or friction. After all, there are institutions that stand to benefit from *suppression* of our creativity. Our job is to *reintroduce* the friction that social media platforms and other digital tools try to take away. Instead of just sliding down the path of least resistance and doomscrolling our lives away, we need to be constantly selecting just what's best for us, using it with a positive outlook, and creatively tailoring it to suit our needs.

The quote by Harold Kushner that opens this conclusion comes from his book *Overcoming Life's Disappointments*. In it, he helps the reader cope with specific issues they might be facing, including relationship difficulties, a death in the family, or financial troubles. Kushner's central message, as embodied by the quote, is that people need to face difficulties not only *to cope* with those challenges but also *to become even stronger*.

There's no question that today's social media environment supplies us with plenty of adversity. But *active reflection*—and making the conscious choice to be more selective, positive, and creative online—is like exercise. It can *strengthen* us, reduce our mental and emotional "flabbiness," and prepare us to handle the challenges that the future will inevitably bring.

The techniques involved in being selective, positive, and creative can apply to the real world as much as the digital world. By doing the hard work of figuring out how we're going to optimize our use of today's social media, we're not just improving our digital lives—we're improving our lives. We may end up not only savvy on social media but also empowered, optimistic, and hopeful in everything we do.

Acknowledgments

For their excellent help with all aspects of the production process, I am grateful to the Chronicle Prism team, which includes Beth Weber, Cecilia Santini, Brooke Johnson, and Tera Killip. I'm also grateful for Jenn Jensen and the team at Wunderkind for their assistance with publicity. Thanks, too, to Mark Tauber and Cara Bedick for believing in me. But I am most indebted to my editor Eva Avery and my copyeditor Jeff Campbell, whose meticulous, thoughtful, and wise editing has been much appreciated and has certainly improved the final product.

I am extremely fortunate to have Lucy Cleland from Kneerim & Williams as my literary agent. I never realized how many different roles must be all rolled up into one to make a great agent—and Lucy is excellent at all of them.

This book would not have been possible without my remarkable research team—the Center for Research on Media, Technology, and Health at the University of Pittsburgh—which had its golden decade from 2009 to 2019. During that time, we amassed hundreds of peer-reviewed papers and presentations that have now been cited over ten thousand times in the scholarly literature. People frequently ask me how we achieved what we did with few resources, and the answer is that it was an organic outpouring from our passion for our work and our compassion for one another. Many faculty members, staff members, students, and other friends enriched the center. However, aside from myself, the central core team members have been—in alphabetical order—Mary Carroll, Kar-Hai Chu, Jason Colditz, César Escobar-Viera, Beth Hoffman, Ariel Shensa, Jaime Sidani, and Michelle Woods.

Most importantly, I owe the deepest gratitude to the family that has stood by me during all the challenging hours of writing, researching, and pouring attention into this book. This of course includes my parents, Karen and Aron, who have been immensely supportive of my endeavors and instilled in me a deep and broad intellectual curiosity. There's also my Fayetteville family, which includes my mother-in-law, Linda, my delightful children, Micah and Sadie, and the love of my life—and my greatest support—Jen.

As for Ellie the dog and Bella and Zoey the guinea pigs, I must acknowledge that they have no idea that there even is a book. And yet, they somehow still know how to make me feel good about it. Go figure.

Notes

Introduction

Page 2: *By the end of the rampage, eleven were*: Robertson C, Mele C, Tavernise S. "11 Killed in Synagogue Massacre; Suspect Charged With 29 Counts." *New York Times*. Published 2018. https://www.nytimes.com/2018/10/27/us/active-shooter-pittsburgh-synagogue-shooting.html.

Page 2: *It quickly became clear from news reports*: Roose K. "On Gab, an Extremist-Friendly Site, Pittsburgh Shooting Suspect Aired His Hatred in Full." *New York Times*. Published 2018. https://www.nytimes.com/2018/10/28/us/gab-robert-bowers-pittsburgh-synagogue-shootings.html.

Page 3: *For example, within days, the Muslim*: Haag M. "Muslim Groups Raise Thousands for Pittsburgh Synagogue Shooting Victims." *New York Times*. Published 2018. https://www.nytimes.com/2018/10/29/us/muslims-raise-money-pittsburgh-synagogue.html.

Page 5: *Further, these changes in our mind can affect*: These details are from Kyrou I, Tsigos C. "Stress Hormones: Physiological Stress and Regulation of Metabolism." *Curr Opin Pharmacol*. 2009;9(6):787-793. doi:10.1016/j.coph.2009.08.007; Reiche EMV, Nunes SOV, Morimoto HK. "Stress, Depression, the Immune System, and Cancer." *Lancet Oncol*. 2004;5(10):617-625. doi:10.1016/S1470-2045(04)01597-9; and McEwen BS. "Protective and Damaging Effects of Stress Mediators." *N Engl J Med*. 1998;338(3):171-179. doi:10.1056/nejm199801153380307.

Page 5: *replaced with the even simpler "MyPlate"*: US Deparment of Agriculture. MyPlate. Published 2021. https://www.myplate.gov.

Page 6: *Instead of a long list of no-nos*: Ibid.

Chapter 1: The Minister of Loneliness

Page 13: *In 2018, the UK prime minister created*: Yeginsu C. "U.K. Appoints a Minister for Loneliness." *New York Times*. Published 2018. https://www.nytimes.com/2018/01/17/world/europe/uk-britain-loneliness.html.

Page 13: *In the United States, 61 percent of people feel lonely*: Cigna. *Loneliness Is at Epidemic Levels in America*. Published 2021. https://www.cigna.com/about-us/newsroom/studies-and-reports/combatting-loneliness.

Page 13: *In many populations, people consider themselves closer*: Howe N. "Millennials and the Loneliness Epidemic." *Forbes*. Published 2019. https://www.forbes.com/sites/neilhowe/2019/05/03/millennials-and-the-loneliness-epidemic.

Page 13: *A 2015 report involving millions of people found*: Holt-Lunstad J, Smith TB, Baker M, Harris T, Stephenson D. "Loneliness and Social Isolation as Risk Factors for Mortality: A Meta-Analytic Review." *Perspect Psychol Sci a J Assoc Psychol Sci*. 2015;10(2):227-237. doi:10.1177/1745691614568352.

Page 14: *Both of these can increase stress hormones*: Reiche, "Stress, Depression, the Immune System."

Page 14: *One-quarter of Americans rate their own mental health*: Cigna, *Loneliness Is at Epidemic*.

Page 14: *In 2016, the* New York Times *reported that the US*: Tavernise S. "U.S. Suicide Rate Surges to a 30-Year High." *New York Times*. Published 2016. https://www.nytimes.com/2016/04/22/health/us-suicide-rate-surges-to-a-30-year-high.html.

Page 14: *Less than a year later, the World Health Organization*: World Health Organization. "'Depression: Let's Talk' Says WHO, as Depression Tops List of Causes of Ill Health." Published 2017. https://www.who.int/news/item/30-03-2017--depression-let-s-talk-says-who-as-depression-tops-list-of-causes-of-ill-health.

Page 14: *Studies have shown that about 20 percent of Americans*: Hasin DS, Sarvet AL, Meyers JL, Saha TD, Ruan WJ, Stohl M, Grant BF. "Epidemiology of Adult DSM-5 Major Depressive Disorder and Its Specifiers in the United States." *JAMA Psychiatry*. 2018;75(4):336-346. doi:10.1001/jamapsychiatry.2017.4602.

Page 14: *For instance, a 2017 study showed that*: Wang J, Wu X, Lai W, Long E, Zhang X, Li W, Zhu Y, Chen C, Zhong X, Liu Z, Wang D, Lin H. "Prevalence of Depression and Depressive Symptoms among Outpatients: A Systematic Review and Meta-Analysis." *BMJ Open*. 2017;7(8):e017173. doi:10.1136/bmjopen-2017-017173.

Page 15: *Emotional conditions like depression and anxiety impact*: Details in this paragraph are from Bautista LE, Vera-Cala LM, Colombo C, Smith P. "Symptoms of Depression and Anxiety and Adherence to Antihypertensive Medication." *Am J Hypertens*. 2012;25(4):505-511. doi:10.1038/ajh.2011.256; O'Neil A, Sanderson K, Oldenburg B. "Depression as a Predictor of Work Resumption Following Myocardial Infarction (MI): A Review of Recent Research Evidence." *Health Qual Life Outcomes*. 2010;8:95. doi:10.1186/1477-7525-8-95; Anker JJ, Kushner MG. "Co-Occurring Alcohol Use Disorder and Anxiety: Bridging Psychiatric, Psychological, and Neurobiological Perspectives." *Alcohol Res*. 2019;40(1). doi:10.35946/arcr.v40.1.03; and Fluharty M, Taylor AE, Grabski M, Munafò MR. "The Association of Cigarette Smoking with Depression and Anxiety: A Systematic Review." *Nicotine Tob Res*. 2017;19(1):3-13. doi:10.1093/ntr/ntw140.

Page 15: *The term* social media *only originated around*: Kemp S. *Digital 2020.* We Are Social. Published 2020. https://wearesocial.com/blog/2020/01/digital-2020-3-8-billion-people-use-social-media.

Page 16: *Today, the most commonly used social media platform*: Ibid.

Page 16: *In January 2018, TikTok had about fifty million users*: Sherman A. "TikTok Reveals Detailed User Numbers for the First Time." CNBC. Published 2020. https://www.cnbc.com/2020/08/24/tiktok-reveals-us-global-user-growth-numbers-for-first-time.html.

Page 16: *Most young adults spend two to four hours per day*: Henderson G. "How Much Time Does the Average Person Spend on Social Media?" Digital Marketing. Published 2020. https://www.digitalmarketing.org/blog/how-much-time-does-the-average-person-spend-on-social-media.

Page 16: *But the fastest-growing population of users*: Pew Research Center. "Social Media Fact Sheet." Published 2019. https://www.pewresearch.org/internet/fact-sheet/social-media.

Chapter 2: Goldilocks Was Nowhere to Be Found

Page 18: *We conducted a national study with about two thousand*: Lin LY, Sidani JE, Shensa A, Radovic A, Miller E, Colditz JB, Hoffman BL, Giles LM, Primack BA. "Association Between Social Media Use and Depression among U.S. Young Adults." *Depress Anxiety*. 2016;33(4):323-331. doi:10.1002/da.22466.

Page 20: *We studied a thousand young adults who were*: Primack BA, Shensa A, Sidani JE, Escobar-Viera CG, Fine MJ. "Temporal Associations Between Social Media Use and Depression." *Am J Prev Med.* 2021;60(2):179-188. doi:10.1016/j.amepre.2020.09.014.

Page 21: *To test this, we started with a mix of people who were*: Shensa A, Sidani JE, Escobar-Viera CG, Switzer GE, Primack BA, Choukas-Bradley S. "Emotional Support from Social Media and Face-to-Face Relationships: Associations with Depression Risk among Young Adults." *J Affect Disord.* 2020;260:38-44. doi:10.1016/j.jad.2019.08.092.

Chapter 3: A Social-Like Experience

Page 22: *In his book* In Defense of Food, *Michael Pollan*: Pollan M. *In Defense of Food: An Eater's Manifesto*. Penguin Press; 2008.

Page 24: *I looked all of this up to confirm what she*: Moolbrock C. "Can Guinea Pigs Die of Loneliness?" Guinea Pig Tube. Published 2020. https://www.guineapigtube.com/can-guinea-pigs-die-of-loneliness.

Page 25: *My colleague Ariel Shensa and I recently published a study*: Shensa. "Emotional Support from Social Media."

Chapter 4: More Powerful Than Advertising

Page 28: *For example, chimpanzees will happily accept just about*: Hopper LM, Lambeth SP, Schapiro SJ, Brosnan SF. "Social Comparison Mediates Chimpanzees' Responses to Loss, Not Frustration." *Anim Cogn.* 2014;17(6):1303-1311. doi:10.1007/s10071-014-0765-9.

Page 28: *In a 2017 French study, baboons were given*: Dumas F, Fagot J, Davranche K, Claidière N. "Other Better Versus Self Better in Baboons: An Evolutionary Approach of Social Comparison." *Proceedings Biol Sci.* 2017;284(1855). doi:10.1098/rspb.2017.0248.

Page 29: *Even guppies—yes, the fish!—compare themselves*: Gasparini C, Serena G, Pilastro A. "Do Unattractive Friends Make You Look Better? Context-Dependent Male Mating Preferences in the Guppy." *Proceedings Biol Sci.* 2013;280(1756):20123072. doi:10.1098/rspb.2012.3072.

Page 29: *Like baboons and guppies, we tend to compare:* Bandura A, Jourden FJ. "Self-Regulatory Mechanisms Governing the Impact of Social Comparison on Complex Decision Making." *J Pers Soc Psychol.* 1991;60(6):941-951. doi:10.1037/0022-3514.60.6.941.

Page 30: *Worldwide advertising spending is now over half*: Guttmann A. "Advertising Media Owners Revenue Worldwide from 2012 to 2024." Statista. Published 2021. https://www.statista.com/statistics/236943/global-advertising-spending.

Page 31: *and this has been supported by research*: Gibbons FX. "Social Comparison and Depression: Company's Effect on Misery." *J Pers Soc Psychol.* 1986;51(1):140-148. doi:10.1037/0022-3514.51.1.140.

Page 31: *Scientific studies have repeatedly shown that chimps*: de Waal F. *Chimpanzee Politics: Power and Sex Among Apes.* Johns Hopkins University Press; 2007.

Chapter 5: The Mean World Syndrome

Page 33: *Our favorite medical dramas, the ones we can't seem*: Primack BA, Roberts T, Fine MJ, Dillman Carpentier FR, Rice KR, Barnato AE. "ER vs. ED: A Comparison of Televised and Real-Life Emergency Medicine." *J Emerg Med.* 2012;43(6). doi:10.1016/j.jemermed.2011.11.002.

Page 35: *a study published in the* Journal of the American Medical: Schlenger WE, Caddell JM, Ebert L, Jordan BK, Rourke KM, Wilson D, Thalji L, Dennis JM, Fairbank JA, Kulka RA. "Psychological Reactions to Terrorist Attacks: Findings from the National Study of Americans' Reactions to September 11." *J Am Med Assoc.* 2002;288(5):581-588.

Page 37: *though the scale of the panic has since been debated*: Hayes J, Battles K. "Exchange and Interconnection in US Network Radio: A Reinterpretation of the 1938 *War of the Worlds* Broadcast." *Int Stud Broadcast Audio Media.* 2011;9(1):51-62. https://doi.org/10.1386/rjao.9.1.51_1.

Page 37: *George Gerbner, a communications scholar who*: Gerbner G. "The 'Mainstreaming' of America: Violence Profile No. 11." *J Commun.* 1980:30(3):10–29. doi:10.1111/j.1460-2466.1980.tb01987.x.

Page 37: *It's estimated that the average youth before turning*: Garbarino J, Bradshaw CP, Vorrasi JA. "Mitigating the Effects of Gun Violence on Children and Youth." *Future Child.* 2002;12(2):73-85. doi:10.2307/1602739.

Page 38: *currently the average time we spend on our phones*: Henderson. "How Much Time Does."

Page 39: *In 2012, the amount of time we spent on social*: Clement J. "Daily Time Spent on Social Networking by Internet Users Worldwide from 2012 to 2019." Statista. Published 2020. https://www.statista.com/statistics/433871/daily-social-media-usage-worldwide.

Page 39: *According to a 2019 study by global tech company*: Asurion. "Americans Check Their Phones 96 Times a Day." Published 2019. https://www.asurion.com/about/press-releases/americans-check-their-phones-96-times-a-day.

Chapter 6: The Good Stuff

Page 41: *As of 2020, people move an average of about*: U.S. Census Bureau. "Calculating Migration Expectancy Using ACS Data." https://www.census.gov/topics/population/migration/guidance/calculating-migration-expectancy.html.

Page 42: *But consider the fact that, by about 2065, there are expected*: Öhman CJ, Watson D. "Are the Dead Taking Over Facebook? A Big Data Approach to the Future of Death Online." *Big Data Soc.* 2019;6(1):2053951719842540. doi:10.1177/2053951719842540.

Page 42: *My colleague Beth Hoffman and I recently published a paper*: Hoffman BL, Shensa A, Escobar-Viera CG, Sidani JE, Miller E, Primack BA. "'Their Page Is Still Up': Social Media and Coping with Loss." *J Loss Trauma.* Published online September 21, 2020:1-18. doi:10.1080/15325024.2020.1820227.

Page 43: *In just one year—from 2018 to 2019—the number*: National Travel and Tourism Office. "U.S. Citizen International Outbound Travel Up Six Percent in 2018." https://travel.trade.gov/tinews/archive/tinews2019/20190402.asp.

Page 44: *About thirty million people—the same number of*: National Human Genome Research Institute. "Rare Diseases FAQ." Published 2020. https://www.genome.gov/FAQ/Rare-Diseases.

Page 44: *Now, for example, there is an active Facebook group for*: National Organization for Rare Disorders, Inc. (NORD). Published 2021. https://www.facebook.com/NationalOrganizationforRareDisorders.

Page 46: *Take the example of the "ice bucket challenge"*: Ice Bucket Challenge. Published 2021. https://en.wikipedia.org/wiki/Ice_Bucket_Challenge.

Page 48: *In 2018, for example, the tobacco industry spent*: Centers for Disease Control and Prevention. Tobacco Industry Marketing. Published 2020. https://www.cdc.gov/tobacco/data_statistics/fact_sheets/tobacco_industry/marketing/index.htm.

Page 48: *In October 2017, actress Alyssa Milano posted on Twitter*: Pflum M. "A Year Ago, Alyssa Milano Started a Conversation about #MeToo. These Women Replied." NBC News. Published 2018. https://www.nbcnews.com/news/us-news/year-ago-alyssa-milano-started-conversation-about-metoo-these-women-n920246.

Page 48: *Starbucks claims that they offer over eighty-seven thousand*: Beck J. "Fancy Starbucks Drinks and the Special Snowflakes Who Order Them." *The Atlantic*. Published 2016. https://www.theatlantic.com/health/archive/2016/05/food-customization-america/482073.

Chapter 7: Why Yo-Yo Tech Diets Don't Work

Page 51: *For example, studies show that FOMO is an important*: Riordan BC, Flett JAM, Hunter JA, Scarf D, Conner TS. "Fear of Missing Out (FoMO): The Relationship between FoMO, Alcohol Use, and Alcohol-Related Consequences in College Students." *Ann Neurosci Psychol*. 2015;2(7):1-7.

Page 52: *As the "father of sociobiology" E. O. Wilson noted, people*: Wilson EO. *On Human Nature*. Harvard University Press; 1978.

Page 52: *When animals cycle through periods of scarcity*: Cifani C, Di Bonaventura MVM, Ciccocioppo R, Massi M. "Binge Eating in Female Rats Induced by Yo-Yo Dieting and Stress BT." In: *Animal Models of Eating Disorders*. Avena NM, ed. Humana Press; 2013:27-49. doi:10.1007/978-1-62703-104-2_3.

Page 52: *With people, problematic chemical changes happen*: Dulloo AG, Montani J-P. "Pathways from Dieting to Weight Regain, to Obesity and to the Metabolic Syndrome: An Overview." *Obes Rev an Off J Int Assoc Study Obes*. 2015;16 Suppl 1:1-6. doi:10.1111/obr.12250.

Page 52: *This style of dieting also increases the risk of*: Thorpe M. "10 Solid Reasons Why Yo-Yo Dieting Is Bad for You." Healthline. Published 2017. https://www.healthline.com/nutrition/yo-yo-dieting.

Chapter 8: The Social Media "Food Pyramid"

Page 56: *In a major study of nearly a hundred thousand women*: Tindle HA, Chang Y-F, Kuller LH, Manson JE, Robinson JG, Rosal MC, Siegle GJ, Matthews KA. "Optimism, Cynical Hostility, and Incident Coronary Heart Disease and Mortality in the Women's Health Initiative." *Circulation*. 2009;120(8):656-662. doi:10.1161/CIRCULATIONAHA.108.827642.

Page 57: *My colleagues and I have found that people with different*: Whaite EO, Shensa A, Sidani JE, Colditz JB, Primack BA. "Social Media Use, Personality Characteristics, and Social Isolation among Young Adults in the United States." *Pers Individ Dif*. 2018;124:45-50. doi:10.1016/j.paid.2017.10.030.

Chapter 9: Select Your Time

Page 63: *However, they think long and hard about these*: Orlowski J, dir. *The Social Dilemma*; 2020.

Page 65: *The longitudinal study my research team and I conducted*: Primack. "Temporal Associations Between Social."

Chapter 10: Select Your Frequency

Page 66: *In* The Distracted Mind, *neuroscientist Adam Gazzaley*: Gazzaley A, Rosen LD. *The Distracted Mind: Ancient Brains in a High-Tech World*. The MIT Press; 2016.

Page 68: *Another reason that frequent use might affect us more*: Firth J, Torous J, Stubbs B, Firth JA, Steiner GZ, Smith L, Alvarez-Jimenez M, Gleeson J, Vancampfort D, Armitage CJ, Sarris J. "The 'Online Brain': How the Internet May Be Changing Our Cognition." *World Psychiatry*. 2019;18(2):119-129. doi:10.1002/wps.20617.

Page 68: *While the process of encoding memories continues*: Norris D. "Short-Term Memory and Long-Term Memory Are Still Different." *Psychol Bull*. 2017;143(9):992-1009. doi:10.1037/bul0000108.

Page 68: *One of the times consolidation happens is during*: Klinzing JG, Niethard N, Born J. "Mechanisms of Systems Memory Consolidation During Sleep." *Nat Neurosci*. 2019;22(10):1598-1610. doi:10.1038/s41593-019-0467-3.

Page 69: *In 2014, a group of Harvard neuroscientists published*: Gregory MD, Agam Y, Selvadurai C, Nagy A, Vangel M, Tucker M, Robertson EM, Stickgold R, Manoach DS. "Resting State Connectivity Immediately Following Learning Correlates with Subsequent Sleep-Dependent Enhancement of Motor Task Performance." *Neuroimage*. 2014;102 Pt 2(0 2):666-673. doi:10.1016/j.neuroimage.2014.08.044.

Page 70: *For example, in* The 4-Hour Workweek, *Tim Ferriss advocates*: Ferriss T. *The 4-Hour Workweek: Escape 9–5, Live Anywhere, and Join the New Rich*. Crown Publishing Group; 2009.

Chapter 11: Select Your Platforms

Page 72: *To solve this question, my research team and I conducted an important study:* Primack BA, Shensa A, Escobar-Viera CG, Barrett EL, Sidani JE, Colditz JB, James AE. "Use of Multiple Social Media Platforms and Symptoms of Depression and Anxiety: A Nationally-Representative Study among U.S. Young Adults." *Comput Human Behav*. 2017;69. doi:10.1016/j.chb.2016.11.013.

Page 73: *In the documentary* The Social Dilemma, *former insiders*: Orlowski. *Social Dilemma*.

Page 73: *Virtual-reality pioneer Jaron Lanier wrote a book called*: Lanier J. *Ten Arguments for Deleting Your Social Media Accounts Right Now*. Henry Holt and Company; 2018.

Chapter 12: Select Something Else Before Bed

Page 76: *In 2005, the American Association for the Advancement of Science*: American Association for the Advancement of Science. "So Much More to Know." *Science*. 2005;309(5731):78-102. doi:10.1126/science.309.5731.78b.

Page 76: *Too little sleep is linked to heart problems*: These details come from Watson S, Cherney K. "The Effects of Sleep Deprivation on Your Body." *Healthline*. Published 2020. https://www.healthline.com/health/sleep-deprivation/effects-on-body; Pilcher JJ, Huffcutt AI. "Effects of Sleep Deprivation on Performance: A Meta-Analysis." *Sleep*. 1996;19(4):318-326. doi:10.1093/sleep/19.4.318; and Orzeł-Gryglewska J. "Consequences of Sleep Deprivation." *Int J Occup Med Environ Health*. 2010;23(1):95-114. doi:10.2478/v10001-010-0004-9.

Page 76: *Interestingly, though,* too much sleep *is also linked to dying*: Wang C, Bangdiwala SI, Rangarajan S, et al. "Association of estimated sleep duration and naps with mortality and cardiovascular events: a study of 116,632 people from 21 countries." *Eur Heart J*. 2019;40(20):1620-1629. doi:10.1093/eurheartj/ehy695.

Page 77: *About two-thirds of us struggle with sleep*: *Consumer Reports*. "Why Americans Can't Sleep." Published 2016. https://www.consumerreports.org/sleep/why-americans-cant-sleep.

Page 77: *For example, compared with people who don't have*: Taylor DJ, Lichstein KL, Durrence HH, Reidel BW, Bush AJ. "Epidemiology of Insomnia, Depression, and Anxiety." *Sleep*. 2005;28(11):1457-1464. doi:10.1093/sleep/28.11.1457.

Page 77: *In our first study, led by sleep psychologist*: Levenson JC, Shensa A, Sidani JE, Colditz JB, Primack BA. "The Association Between Social Media Use and Sleep Disturbance among Young Adults." *Prev Med.* 2016;85:36-41. doi:10.1016/j.ypmed.2016.01.001.

Page 78: *To answer that question, we conducted another study*: Levenson JC, Shensa A, Sidani JE, Colditz JB, Primack BA. "Social Media Use Before Bed and Sleep Disturbance among Young Adults in the United States: A Nationally Representative Study." *Sleep.* 2017;40(9):1-7. doi:10.1093/sleep/zsx113.

Page 79: *There's no good evidence that:* Nagare R, Plitnick B, Figueiro M. "Does the iPad Night Shift Mode Reduce Melatonin Suppression?" *Light Res Technol.* 2019;51(3):373-383. doi:10.1177/1477153517748189.

Page 80: *Studies suggest that keeping a gratitude journal*: Singh M. "If You Feel Thankful, Write It Down. It's Good for Your Health." National Public Radio. Published 2018. https://www.npr.org/sections/health-shots/2018/12/24/678232331/if-you-feel-thankful-write-it-down-its-good-for-your-health.

Chapter 13: Select People You Know Well

Page 82: *My colleague Ariel Shensa and I looked at this*: Shensa A, Sidani JE, Escobar-Viera CG, Chu KH, Bowman ND, Knight JM, Primack BA. "Real-Life Closeness of Social Media Contacts and Depressive Symptoms among University Students." *J Am Coll Heal.* 2018;66(8):747-753. doi:10.1080/07448481.2018.1440575.

Page 82: *The relationship between having more non-face-to-face friends*: Putnam RD. *Bowling Alone: The Collapse and Revival of American Community.* Simon & Schuster; 2000.

Page 83: *In that same study, we also asked participants*: Shensa. "Real-Life Closeness of Social."

Page 84: *Author Marie Kondo instructs people to hold up every*: Kondo M. *The Life-Changing Magic of Tidying Up: The Japanese Art of Decluttering and Organizing.* Ten Speed Press; 2014.

Chapter 14: Select Your Digital Holidays

Page 87: *But a pretty good body of research shows*: Still J. "5 Reasons to Wake Up Early, Even If You Hate It." *Business Insider*. Published 2019. https://www.businessinsider.com/reasons-to-wake-up-early-2019-5.

Page 87: *In 2016, a Danish researcher conducted an experiment*: Tromholt M. "The Facebook Experiment: Quitting Facebook Leads to Higher Levels of Well-Being." *Cyberpsychology, Behav Soc Netw.* 2016;19(11):661-666. doi:10.1089/cyber.2016.0259.

Page 88: *This movement began in 2017, when there were two*: Brown J. "Offline October: Littleton Students Organize Social Media Blackout after Teen Suicides." *Denver Post*. Published 2017. https://www.denverpost.com/2017/10/13/offline-october-student-social-media-blackout.

Page 89: *One of the most compelling speakers I heard was Tiffany Shlain*: Tiffany Shlain. Internet Movie Database. https://www.imdb.com/name/nm1286067.

Page 89: *This idea is echoed by J. Dana Trent, a Baptist minister*: Trent JD. *For Sabbath's Sake: Embracing Your Need for Rest, Worship, and Community*. Upper Room Books; 2017.

Part 3: Be Positive

Page 92: *Various studies suggest that optimism may increase and improve work*: These studies include the following: Gielan M. "The Financial Upside of Being an Optimist." *Harvard Business Review*. Published 2019. https://hbr.org/2019/03/the-financial-upside-of-being-an-optimist; Andersson G. "The Benefits of Optimism: A Meta-Analytic Review of the Life Orientation Test." *Pers Individ Dif*. 1996;21(5):719-725. doi:https://doi.org/10.1016/0191-8869(96)00118-3; Shapira LB, Mongrain M. "The Benefits of Self-Compassion and Optimism Exercises for Individuals Vulnerable to Depression." *J Posit Psychol*. 2010;5(5):377-389. doi:10.1080/17439760.2010.516763; Gallagher MW, Lopez SJ, Pressman SD. "Optimism Is Universal: Exploring the Presence and Benefits of Optimism in a Representative Sample of the World." *J Pers*. 2013;81(5):429-440. doi:https://doi.org/10.1111/jopy.12026; and Tindle. "Optimism, Cynical Hostility."

Page 92: *after Martin Seligman and Mihaly Csikszentmihalyi published a seminal*: Seligman MEP, Csikszentmihalyi M. "Positive Psychology: An Introduction." *Am Psychol*. 2000;55(1):5-14.

Page 93: *Martin Seligman came up with the acronym PERMA*: Kern ML, Waters LE, Adler A, White MA. "A Multidimensional Approach to Measuring Well-Being in Students: Application of the PERMA Framework." *J Posit Psychol*. 2015;10(3):262-271. doi:10.1080/17439760.2014.936962; and Slavin SJ, Schindler D, Chibnall JT, Fendell G, Shoss M. "PERMA: A Model for Institutional Leadership and Culture Change." *Acad Med*. 2012;87(11). https://journals.lww.com/academicmedicine/Fulltext/2012/11000/PERMA__A_Model_for_Institutional_Leadership_and.25.aspx.

Chapter 15: The Power of Negativity

Page 97: *To better understand the effects of negativity, my team*: Primack BA, Bisbey MA, Shensa A, Bowman ND, Karim SA, Knight JM, Sidani JE. "The Association Between Valence of Social Media Experiences and Depressive Symptoms." *Depress Anxiety*. 2018; 35(8):784-794. doi:10.1002/da.22779.

Page 97: *In our study, we did find that an increase in positive*: Ibid.

Page 97: *To make things worse, in different studies we conducted*: Primack BA, Karim SA, Shensa A, Bowman N, Knight J, Sidani JE. "Positive and Negative Experiences on Social Media and Perceived Social Isolation." *Am J Heal Promot*. 2019; 33(6):859-868. doi:10.1177/0890117118824196.

Chapter 16: Find Your Negativity Threshold

Page 101: *The average Facebook user has about 350 friends*: Osman M. "Wild and Interesting Facebook Statistics and Facts (2021)." Kinsta. Published 2021. https://kinsta.com/blog/facebook-statistics.

Page 105: *In a remarkable study conducted at the University of California*: Mullin CR, Linz D. "Desensitization and Resensitization to Violence Against Women: Effects of Exposure to Sexually Violent Films on Judgments of Domestic Violence Victims." *J Pers Soc Psychol*. 1995;69(3):449-459. doi:10.1037//0022-3514.69.3.449.

Chapter 17: Keep Your Feather Pillows Intact

Page 106: *There are many different versions of this story, including*: Brombacher S. "A Pillow Full of Feathers." Chabad.org. Published 2009. https://www.chabad.org/library/article_cdo/aid/812861/jewish/A-Pillow-Full-of-Feathers.htm.

Page 107: *New artificial intelligence programs can make it look*: Vincent J. "New AI Research Makes It Easier to Create Fake Footage of Someone Speaking." *Verge*. Published 2017. https://www.theverge.com/2017/7/12/15957844/ai-fake-video-audio-speech-obama.

Page 107: *And published research suggests that untrue rumors may spread*: Zubiaga A, Liakata M, Procter R, Wong Sak Hoi G, Tolmie P. "Analysing How People Orient to and Spread Rumours in Social Media by Looking at Conversational Threads." Masuda N, ed. *PLoS One*. 2016;11(3):e0150989. doi:10.1371/journal.pone.0150989.

Page 109: *They often use Facebook to post benign content*: https://ink-co.com/insights/5-ways-snapchat-changing-social-media.

Chapter 18: Balance Confidence and Humility

Page 111: *Moore was born into an Irish Catholic family*: "Thomas Moore." poetryfoundation.org. Published 2021. https://www.poetryfoundation.org/poets/thomas-moore.

Page 111: *Weil's situation was also complex*: Petrement S. *Simone Weil: A Life*. Schocken; 1988.

Page 113: *A 2018 report suggested that an individual with over a million*: Liber C. "How and Why Do Influencers Make So Much Money? The Head of an Influencer Agency Explains." *Vox*. Published 2018. https://www.vox.com/the-goods/2018/11/28/18116875/influencer-marketing-social-media-engagement-instagram-youtube.

Chapter 19: Be Positive with a Vengeance

Page 116: *The answer comes from Viktor Frankl, one of the most*: Frankl VE. *Man's Search for Meaning*. Beacon Press; 2006.

Page 117: *Another quotation when considering how to actively pursue*: Towles A. *Rules of Civility*. Penguin Books; 2011.

Page 118: *Platforms are often optimized . . . similar algorithms and processes are likely at work*: Orlowski. *Social Dilemma.*

Chapter 20: Dance Like the Whole Universe Is Watching

Page 121: *In early 2019, a man was recorded on a stranger's phone*: Vaidyanathan V. "Man Drags Daughter Through Airport by Her Jacket Hood." *International Business Times*. Published 2019. https://www.ibtimes.com/watch-man-drags-daughter-through-airport-her-jacket-hood-2750455.

Page 122: *For example, just about every university vets future employees*: Moody J. "Why Colleges Look at Students' Social Media." *U.S. News and World Report*. Published 2019. https://www.usnews.com/education/best-colleges/articles/2019-08-22/why-colleges-look-at-students-social-media-accounts; and Wilkinson S. "This Is How Employers Screen Your Old Social Media Posts." Vice.com. Published 2019. https://www.vice.com/en/article/evy8vz/social-media-vetting-process-future-employers.

Part 4: Be Creative

Page 129: *Social media marketers have been known to use the strategy*: Orlowski. *Social Dilemma*.

Page 130: *For example, one major tobacco company categorized*: Braun S, Mejia R, Ling PM, Pérez-Stable EJ. "Tobacco Industry Targeting Youth in Argentina." *Tob Control*. 2008;17(2):111-117. doi:10.1136/tc.2006.018481.

Page 130: *Public health professionals also use these same techniques*: Berg CJ, Haardörfer R, Getachew B, Johnston T, Foster B, Windle M. "Fighting Fire with Fire: Using Industry Market Research to Identify Young Adults at Risk for Alternative Tobacco Product and Other Substance Use." *Soc Mar Q*. 2017;23(4):302-319. doi:10.1177/1524500417718533; de Toro-Martín J, Arsenault BJ, Després J-P, Vohl M-C. "Precision Nutrition: A Review of Personalized Nutritional Approaches for the Prevention and Management of Metabolic Syndrome." *Nutrients*. 2017;9(8). doi:10.3390/nu9080913; and

Kreuter M, Farrell D, Olevitch L, Brennan L. *Tailoring Health Messages: Customizing Communication With Computer Technology*. Routledge; 1999.

Page 132: *and it's based on established personality research*: Gosling SD, Rentfrow PJ, Swann WB. "A Very Brief Measure of the Big-Five Personality Domains." *J Res Pers*. 2003;37(6):504-528. doi:10.1016/S0092-6566(03)00046-1.

Chapter 21: Consciously Conscientious

Page 134: *My research team and I explored this question by testing*: Whaite. "Social Media Use, Personality."

Chapter 22: The Double-Edged Sword of Agreeableness

Page 139: *We found that more social media use translated into more*: Whaite. "Social Media Use, Personality."

Chapter 23: Embrace Your Neurotic Side Before *and* After Posting

Page 142: *Being neurotic may actually come along with a survival advantage*: Allen NB, Badcock PBT. "Darwinian Models of Depression: A Review of Evolutionary Accounts of Mood and Mood Disorders." *Prog Neuropsychopharmacol Biol Psychiatry*. 2006;30(5):815-826. doi:10.1016/j.pnpbp.2006.01.007.

Page 143: *In our studies, we found that—without even considering*: Whaite. "Social Media Use, Personality."

Chapter 25: Extroversion: Which Way Do You Turn?

Page 150: *Some of these books include Susan Cain's*: Cain S. *Quiet: The Power of Introverts in a World That Can't Stop Talking*. Crown Publishing Group; 2013; Pollard M, Lewis D. *The Introvert's Edge: How the Quiet and Shy Can Outsell Anyone*. HarperCollins Leadership; 2018; Laney MO. *The Introvert Advantage: How Quiet People Can Thrive in an Extrovert World*. Workman Publishing Company; 2002; and Granneman J. *The Secret Lives of Introverts: Inside Our Hidden World*. Skyhorse Publishing; 2017.

Page 151: *Mark Zuckerberg, the longtime chief executive officer*: "5 Mega-Successful Entrepreneurs Who Are Introverts." *Entrepreneur*. Published 2017. https://www.entrepreneur.com/article/286611.

Page 152: *A good example of the difference between objective*: Turnbull C. *The Forest People: A Study of the Pygmies of the Congo*. Simon and Schuster; 1962.

Page 155: *In Virginia Woolf's masterpiece* A Room: Woolf V. *A Room of One's Own*. Hogarth Press; 1929.

Chapter 26: Program or Be Programmed

Page 156: *Officially, there's no such thing as addiction*: Pies R. "Should DSM-V Designate 'Internet Addiction' a Mental Disorder?" *Psychiatry (Edgmont)*. 2009;6(2):31-37. https://pubmed.ncbi.nlm.nih.gov/19724746.

Page 156: *In 2019, an eleven-year-old Indiana boy shot his father*: Becker L. "State Trooper Shot by 11-Year-Old Son Because Video Games Were Taken Away." *WSBT News*. Published 2019. https://wsbt.com/news/local/docs-state-trooper-shot-by-11-year-old-son-because-his-video-games-were-taken-away.

Page 157: *In another incident, a sixteen-year-old boy shot both*: "Daniel Petric Killed Mother, Shot Father Because They Took Halo 3 Video Game, Prosecutors Say." *Cleveland Plain Dealer*. Published 2008. https://www.cleveland.com/metro/2008/12/boy_killed_mom_and_shot_dad_ov.html.

Page 158: *This led to an insightful study that my colleague Ariel*: Shensa A, Escobar-Viera CG, Sidani JE, Bowman ND, Marshal MP, Primack BA. "Problematic Social Media Use and Depressive Symptoms among U.S. Young Adults: A Nationally-Representative Study." *Soc Sci Med*. 2017;182 150-157. doi:10.1016/j.socscimed.2017.03.061.

Page 158: *In his 2010 book* Program or Be Programmed: Rushkoff D. *Program or Be Programmed: Ten Commands for a Digital Age*. OR Books; 2011.

Page 159: *Psychiatrist César Escobar-Viera and I conducted a recent*: Escobar-Viera CG, Shensa A, Bowman ND, Sidani JE, Knight J, James AE, Primack BA. "Passive and Active Social Media Use and Depressive Symptoms among United States Adults." *Cyberpsychology, Behav Soc Netw*. 2018;21(7):437-443. doi:10.1089/cyber.2017.0668.

Page 158: *There is an established scale used to measure "problematic social media use."*: Andreassen C, Torsheim T, Brunborg G, Pallesen S. "Development of a Facebook Addiction Scale." *Psychol Rep*. 2012;110:501-17. doi:10.2466/02.09.18.PR0.110.2.501-517.

Chapter 27: Develop Social Media Literacy

Page 162: *They are happy because they are responding to*: Mai L-W, Schoeller G. "Emotions, Attitudes and Memorability Associated with TV Commercials." *J Targeting, Meas Anal Mark*. 2009;17(1):55-63. doi:10.1057/jt.2009.1.

Page 163: *Because experiments have shown that humans connect yellow*: van Braam, Hailey. *How Color Affects Appetite in Marketing*. Published online April 24, 2020. Available at: https://www.colorpsychology.org/color-appetite; and Clydesdale FM, Gover R, Fugardi C. "The Effect of Color on Thirst Quenching, Sweetness, Acceptability and Flavor Intensity in Fruit

Punch Flavored Beverages." *J Food Qual*. 1992;15(1):19-38. doi:https://doi.org/10.1111/j.1745-4557.1992.tb00973.x.

Page 163: *McDonald's is the largest distributor of toys in the world*: Schlosser E. *Fast Food Nation: The Dark Side of the All-American Meal*. Houghton Mifflin; 2012.

Page 164: *One Harvard professor suggested that as much as 95 percent*: Zaltman G. *How Customers Think: Essential Insights into the Mind of the Market*. Harvard Business School Press; 2003.

Page 164: *It's suggested by the American Academy of Pediatrics*: Turner KH, Jolls T, Hagerman MS, O'Byrne W, Hicks T, Eisenstock B, Pytash KE. "Developing Digital and Media Literacies in Children and Adolescents." *Pediatrics*. 2017;140(Supplement 2):S122 LP-S126. doi:10.1542/peds.2016-1758P.

Page 165: *Over the years, I developed a mnemonic to make it:* Primack BA, Douglas EL, Land SR, Miller E, Fine MJ. "Comparison of Media Literacy and Usual Education to Prevent Tobacco Use: A Cluster-Randomized Trial." *J Sch Health*. 2014;84(2). doi:10.1111/josh.12130.

Chapter 28: Consider Your VQ

Page 168: *And, in fact, our EQ can in certain contexts be even more*: Deutschendorf H. "Why Emotionally Intelligent People Are More Successful." *Fast Company*. Published 2015. https://www.fastcompany.com/3047455/why-emotionally-intelligent-people-are-more-successful.

Page 168: *One key issue she's concerned about is the portrayal*: Silverman S. "Jameela Jamil Slams Diet Culture and Quick-Fix Weight Loss on Social Media." Health.com. https://www.health.com/celebrities/jameela-jamil-petition-toxic-diet-products.

Page 169: *Studies have shown that those diagnosed with*: Arcelus J, Mitchell AJ, Wales J, Nielsen S. "Mortality Rates in Patients with Anorexia Nervosa and Other Eating Disorders. A Meta-Analysis of 36 Studies." *Arch Gen Psychiatry*. 2011;68(7):724-731. doi:10.1001/archgenpsychiatry.2011.74.

Page 169: *In a classic study conducted in Fiji in the mid-1990s*: Becker AE, Burwell RA, Gilman SE, Herzog DB, Hamburg P. "Eating Behaviours and Attitudes Following Prolonged Exposure to Television among Ethnic Fijian Adolescent Girls." *Br J Psychiatry*. 2002;180:509-514. doi:10.1192/bjp.180.6.509.

Page 170: *Concerned about the effect of social media on eating*: Sidani JE, Shensa A, Hoffman B, Hanmer J, Primack BA. "The Association between Social Media Use and Eating Concerns among US Young Adults." *J Acad Nutr Diet*. 2016;116(9):1465-1472. doi:10.1016/j.jand.2016.03.021.

Page 171: *A few studies have ranked social media sites*: Cohen R, Newton-John T, Slater A. "The Relationship between Facebook and Instagram Appearance-Focused Activities and Body Image Concerns in Young Women." *Body Image*. 2017;23:183-187. doi:10.1016/j.bodyim.2017.10.002; and Engeln R, Loach R, Imundo MN, Zola A. "Compared to Facebook, Instagram Use Causes More Appearance Comparison and Lower Body Satisfaction in College Women." *Body Image*. 2020;34:38-45.

Chapter 29: The Eloquence of Emoji 

Page 172: *For the first time in history, the Oxford English Dictionary's*: Steinmetz K. "Oxford's 2015 Word of the Year Is This Emoji." *Time Magazine*. Published 2015. https://time.com/4114886/oxford-word-of-the-year-2015-emoji.

Chapter 30: Become an Alert Wizard

Page 176: *In 1961, Kurt Vonnegut published a short story*: Vonnegut K. "Harrison Bergeron." *Magazine of Fantasy and Science Fiction*; 1961.

Page 177: *A 2014 study published in the journal* Social: Thornton B, Faires A, Robbins M, Rollins E. "The Mere Presence of a Cell Phone May Be Distracting: Implications for Attention and Task Performance." *Soc Psychol*. 2014;45(6):479-488. doi:10.1027/1864-9335/a000216.

Chapter 31: The Good Stuff Is Often in the Back of the Room

Page 182: *According to the insider documentary* The Social Dilemma: Orlowski. *Social Dilemma*.

Chapter 33: Create Your Social Media Checklist

Page 189: *In 2009, surgeon Atul Gawande published* The Checklist Manifesto: Gawande A. *The Checklist Manifesto: How to Get Things Right*. Picador; 2010.

Part 5: How to Live Now

Page 197: *In 1997, the biggest social media site in the world*: Samur A. "The History of Social Media: 29+ Key Moments." *Hootsuite*. Published 2018. https://blog.hootsuite.com/history-social-media/.

Chapter 34: Work

Page 201: *Of people employed full-time in the United States in a single*: U.S. Department of Labor. *American Time Use Survey*. Published 2020. https://www.bls.gov/news.release/pdf/atus.pdf.

Page 201: *In the 1930s, John Maynard Keynes wrote*: Keynes JM. *Economic Possibilities for Our Grandchildren*. Harcourt Brace; 1932.

Page 201: *A 2014 Gallup poll asked college graduates whether their job:* Riffkin R. "In U.S., 55% of Workers Get Sense of Identity from Their Job: Seven in 10 College Graduates Get a Sense of Identity from Their Job." Gallup Polls. Published 2014. https://news.gallup.com/poll/175400/workers-sense-identity-job.aspx.

Page 202: *Now, according to a CareerBuilder survey, about 70*: Salm L. "70% of Employers Are Snooping Candidates' Social Media Profiles." CareerBuilder. Published 2017. https://www.careerbuilder.com/advice/social-media-survey-2017.

Page 203: *A recent survey showed that—among the college admissions*: Jaschik S. "Rise Seen in Admissions Officers Checking Social Media." Inside Higher Ed. Published 2020. https://www.insidehighered.com/admissions/article/2020/01/13/more-admissions-officers-last-year-check-social-media.

Page 204: *One of my students, Jacquelynn Jones, and I published*: Jones JR, Colditz JB, Shensa A, Sidani JE, Lin LY, Terry MA, Primack BA. "Associations between Internet-Based Professional Social Networking and Emotional Distress." *Cyberpsychology, Behav Soc Netw*. 2016;19(10):601-608. doi:10.1089/cyber.2016.0134.

Chapter 35: Family Life

Page 206: *Ten-year-olds in the United States can name more brands*: "Kids Can Identify Booze But Not U.S. Presidents." *Deseret News*. Published 1988. https://www.deseret.com/1988/9/5/18777399/kids-can-identify-booze-but-not-u-s-presidents.

Page 206: *And aggressive social media campaigns promote Budweiser*: Weggert T. "3 Beer Brands Brewing Great Social Media Campaigns." The Content Strategist. Published 2016. https://contently.com/2016/06/06/3-beer-brands-brewing-great-social-media-campaigns.

Page 206: *Starting in 2017, there was an extensive Bud Light mass*: Flanagan G, Dua T. "Bud Light's 'Dilly Dilly' Just Made a Comeback at the Super Bowl with a Weird Crossover Ad with Game of Thrones—Here's What the Phrase Means." *Business Insider*. Published 2019. https://www.businessinsider.com/bud-light-dilly-dilly-viral-commercial-super-bowls-campaign-2017-12.

Page 207: *Youth are especially vulnerable to these psychological*: Te'eni-Harari T. "Clarifying the Relationship between Involvement Variables and Advertising Effectiveness among Young People." *J Consum Policy*. 2014;37(2):183-203. doi:10.1007/s10603-013-9226-0.

Page 207: *The FBI estimates that a half million sexual predators*: Kempf V. "Stats About Online Predators and Precautions Parents Should Take."

Patch. Published 2012. https://patch.com/massachusetts/sudbury/bp--stats-about-online-predators-and-precautions-parec47b01a336.

Page 207: *It's estimated that about 36 percent of youth feel*: Patchin J. "2019 Cyberbullying Data." Cyberbullying Research Center. https://cyberbullying.org/2019-cyberbullying-data.

Page 209: *Indeed, later studies about DARE's effectiveness showed*: Studies in this paragraph include West SL, O'Neal KK. "Project D.A.R.E. Outcome Effectiveness Revisited." *Am J Public Health*. 2004;94(6):1027-1029. doi:10.2105/AJPH.94.6.1027; and Rosenbaum DP, Hanson GS. "Assessing the Effects of School-Based Drug Education: A Six-Year Multilevel Analysis of Project D.A.R.E." *J Res Crime Delinq*. 1998;35(4):381-412. doi:10.1177/0022427898035004002.

Chapter 36: Tech Is for the Young, Right?

Page 212: *In fact, a recent Pew research study found that 50*: Ausubel J. "Older People Are More Likely to Live Alone in the U.S. Than Elsewhere in the World." Pew Research Center. Published 2020. https://www.pewresearch.org/fact-tank/2020/03/10/older-people-are-more-likely-to-live-alone-in-the-u-s-than-elsewhere-in-the-world.

Page 214: *The number of individuals sixty-five and older is*: Meinert M. "Seniors Will Soon Outnumber Children, but the U.S. Isn't Ready." University of Southern California News. Published 2018. https://news.usc.edu/143675/aging-u-s-population-unique-health-challenges.

Page 214: *It's projected that, by 2030, the percent of*: Gray G, Hayes TO, Strohman A. "The Future of America's Entitlements: What You Need to Know About the Medicare and Social Security Trustees Reports." American Action Forum. Published 2019. https://www.americanactionforum.org/research/the-future-of-americas-entitlements-what-you-need-to-know-about-the-medicare-and-social-security-trustees-reports-2.

Page 215: *At the University of Arkansas, we're facing up to scammers*: "OLLI and SALT Teaming Up to Host Senior Citizen Safety Programs in Fayetteville." University of Arkansas News. Published 2019. https://news.uark.edu/articles/50088/olli-and-salt-teaming-up-to-host-senior-citizen-safety-programs-in-fayetteville.

Chapter 37: The Holidays and Consumerism

Page 217: *According to the National Science Foundation, we have*: Verma P. "Destroy Negativity from Your Mind with This Simple Exercise." *Medium*. Published 2017. https://medium.com/the-mission/a-practical-hack-to-combat-negative-thoughts-in-2-minutes-or-less-cc3d1bddb3af.

Page 218: *The National Retail Federation estimates that holiday sales*: National Retail Federation. "Winter Holiday FAQ." https://nrf.com/insights/holiday-and-seasonal-trends/winter-holidays/winter-holiday-faqs.

Page 219: *Consider the story of the Minneapolis man who angrily strode*: Duhigg C. "How Companies Learn Your Secrets." *New York Times Magazine*. Published 2012. https://www.nytimes.com/2012/02/19/magazine/shopping-habits.html.

Chapter 38: When You're Vulnerable

Page 221: *Several years ago, I read a fascinating book called*: Nesse RM, Williams GC. *Why We Get Sick*. Vintage Books; 1996.

Conclusion

Page 231: *One of my favorite illustrations of this is a classic paper*: Miller GA. "The Magical Number Seven, Plus or Minus Two: Some Limits on Our Capacity for Processing Information." *Psychol Rev*. 1956;63(2):81-97.

Page 234: *The quote by Harold Kushner that opens this conclusion*: Kushner HS. *Overcoming Life's Disappointments: Learning from Moses How to Cope with Frustration*. Anchor Books; 2007.

About the Author

BRIAN A. PRIMACK has a multidisciplinary background spanning the humanities, social sciences, and medicine. After graduating Yale University in 1991 with degrees in English and Mathematics, he taught adolescents in West Africa and studied educational psychology and human development for his first master's degree, which he received from Harvard University. He then graduated first in his class from Emory Medical School and trained in Family Medicine in Pittsburgh. He subsequently received another master's degree and a PhD focused on social and behavioral science from the University of Pittsburgh while on faculty there.

He is an international thought leader around the complex interrelationships between media, technology, and health. He has received over $10 million in research funding to study these topics, and his findings have been published in multiple leading journals in medicine, public health, and social science. These results have been reported in news sources such as *NPR's All Things Considered, NPR's Here and Now*, the *New York Times, The Washington Post, US News and World Report, CBS Sunday Morning with Jane Pauley* . . . and even *Cosmo.*

He currently serves as Dean of the College of Education and Health Professions at the University of Arkansas. In this role, he guides a highly diverse College serving over 5,000 undergraduate and graduate students and 500 faculty and staff members in areas such as K–12 education, nursing, occupational and speech therapy, adult and community education, exercise science, educational policy, counseling, and public health. Dr. Primack also serves as the Henry G. Hotz Endowed Chair in Educational Innovations and a Professor of Public Health and Medicine.

Over the years, he's also worked professionally as an actor, cryptologist (code breaker), musician, physical education teacher, dean of students, busboy, and improvisational comedian.

He currently lives in Fayetteville, Arkansas, with his wife Jen, two teenage children Micah and Sadie, his mother-in-law Linda, Ellie the dog, and Bella and Zoey the guinea pigs.

Index

J

K

L

M

N

O

P

R

S

T

V

W

Y

Z